D1010568

A TERRIBLE THING TO WASTE

Also by Harriet A. Washington

Infectious Madness

Deadly Monopolies

Medical Apartheid

A TERRIBLE THING TO WASTE

ENVIRONMENTAL RACISM AND ITS ASSAULT ON THE AMERICAN MIND

HARRIET A. WASHINGTON

LITTLE, BROWN SPARK
New York Boston London

For Ron DeBose, my husband, in memoriam

Little, Brown Spark
Hachette Book Group
1290 Avenue of the Americas, New York, NY 10104
littlebrownspark.com

First Edition: July 2019

Little Brown Spark is an imprint of Little, Brown and Company, a division of Hachette Book Group, Inc. The Little, Brown Spark name and logo are trademarks of Hachette Book Group, Inc.

ISBN 978-0-316-50943-5
LCCN 2018957423

10 9 8 7 6 5 4 3 2 1

LSC-C

Printed in the United States of America

Contents

Humankind has not woven the web of life. We are but one thread within it. Whatever we do to the web, we do to ourselves.

— Chief Seattle

A TERRIBLE THING TO WASTE

INTRODUCTION

IQ Matters

Nature has color-coded groups of individuals so that statistically reliable predictions of their adaptability to intellectually rewarding and effective lives can easily be made.

— William Shockley, "Models, Mathematics, and the Moral Obligation to Diagnose the Origin of Negro IQ Deficits," *Review of Educational Research*[1]

Long before William Shockley's 1971 assertion that the intelligence of Americans is innate, inherited, and permanently stratified by race, the nineteenth-century scientists known as the American School of Ethnology had trumpeted the same belief, sans statistics. Their assertions were preceded by a long, data-free history of speculation by Europeans on the lower intelligence of Africans and their descendants, speculation that had traditionally supplied a rationale for enslavement.

The question of whether innate differences in intelligence exist between blacks and whites goes back more than a thousand years, to the time when the Moors invaded Europe. Although we focus on European claims that Africans and their descendants are relatively unintelligent, some people of the African diaspora, including the scientifically adept and accomplished Moors, have returned the favor. As Richard E. Nisbett, author of *Intelligence and How to Get It: Why Schools and Cultures Count,* writes, "The Moors speculated that Europeans might be congenitally incapable of abstract thought."[2]

But no Moors practiced science in the United States, and in 1840 the government falsified census data to support enslavement with the spurious claim that slaves enjoyed much better mental health than freedmen. Thereafter, freedom was held to be detrimental to African Americans' mental health.[3]

In 1981, Stephen Jay Gould's *The Mismeasure of Man* documented the long history of painstaking but rigged data collection and analysis enlisted to support the cherished belief in the innate intellectual inferiority of blacks. So, too, have *Even the Rat Was White,* first published in 1976, by the late psychologist Robert V. Guthrie, as well as a slew of more recent works.[4]

Today, the hereditarian agenda still drives fevered debate about intelligence, or to be more precise, about IQ. Accusations of "racism" and "political correctness" fly in reaction to each provocative new publication on race and IQ. The headlines, and too much of the scientific discourse, dwell on the 15-point gap between the average scores of U.S. whites and African Americans.

Much hand-wringing ensues: What can be done? Are interventions futile? In the end, this very public and very political drama drives a false perception of genetics as the chief factor determining IQ. And the sniping between hereditarians and their critics has accomplished nothing except a great deal of harping on the prohibitive expense and futility of intervening to close the gap. After all, hereditarians argue, nothing can be done to correct the "innate," genetically dictated, lower-IQ status of African Americans. In tones of infinite regret, their screeds and interviews insist that devoting resources to closing it would be a Quixotic and irresponsibly expensive task: the racial IQ gap is impervious to treatment and here to stay.

This bleak prospect is not supported by the facts.

Salt of the Earth

Consider that in 1924, the United States closed another 15-point IQ gap with a single, cheap step: we added iodine to salt.[5] The government didn't set out to increase IQ—that was a happy accident, the consequence of a program intended to address nutritional deficiency diseases, chiefly goiters, that are caused by insufficient iodine in the diet.

Before 1920, whether or not your diet contained sufficient iodine depended largely on where your water came from. Seawater, for instance, is rich in iodine, which is fortunate for countries like Japan, which has some of the world's most iodine-rich water, and boasts one of the world's highest average IQs.

But in a place like Michigan, where the natural water supply was largely runoff from iodine-free glaciers, much of the water and soil in which vegetables grew lacked iodine. This deficiency heightened the rates of goiter, an enlarged thyroid that causes an unsightly lump in the neck. Goiters are rarely life-threatening, but some require medication or surgery.[6]

For just a few dollars a ton, adding iodine to salt (in the form of *potassium iodide*) greatly lowered the incidence of deficiency and goiters. But during World War II, when a new crop of Air Force recruits was tested with the Army General Classification Test, scientists noticed a pattern: men from low-iodine areas who showed improved thyroid health also scored substantially higher on the intelligence test. It turns out that iodine deficiency is an unrecognized cause of mental retardation, especially, but not exclusively, in unborn children. Now we know that iodine bolsters healthy brain development because it is an essential component of thyroid hormone, which helps to direct the complex development of the fetal brain.

Yet iodine deficiency remains the world's leading cause of preventable mental retardation. Prenatal iodine deficiency as well as deficiency in adulthood leads to compromised intelligence as measured by IQ testing, especially in the developing nations that report some of the world's lowest IQ scores. In 2007, Kul Gautam, the deputy executive director of UNICEF, announced, "Today over one billion people in the world suffer from iodine deficiency, and 38 million babies born every year are not protected from brain damage due to IDD."[7]

Almost one-third of the world's population has too little iodine in its diet, and the problem isn't limited to developing countries: in Europe, iodized salt is still not the norm. Moreover, the health-conscious trend toward reducing salt intake has lowered iodine intake as well; so has the gourmet penchant for natural sea salt, which does not contain iodine. Within the United States, African Americans suffer the nation's highest level of iodine deficiency. According to the government's National Health and Nutrition Examination Survey (NHANES) data for 2005–2008, African Americans' urinary iodine reading of 137 µg/L is lower than that of Mexican Americans at 174 µg/L and of non-Hispanic whites at 168 µg/L. No separate data are given for Native Americans or other ethnic groups.

Today, nearly a century after iodine's effects on the brain were first discovered, a wealth of contemporary studies warns of many other environmental factors that, like iodine, dramatically affect intelligence and IQ. These range from lead, which recently made national headlines for contaminating the drinking water in Flint, Michigan, to other poisonous metals such as arsenic and mercury, which often accompany lead and commonly poison reservation lands.

Unsurprisingly, the deleterious effects of chemicals in the environment have been most widely reported when they impact white, relatively affluent communities. In the 1980s, for instance, headlines frequently updated us on the struggles of communities like Love Canal in Niagara Falls, New York, where a seventy-acre landfill spewed pollution that threatened the health of hundreds.

But we never read about Anniston.

Saving Anniston

I first heard about Anniston, Alabama, during a chance encounter in the heart of New York City. In 2005, I spent most weekdays at the New York Public Library's iconic Stephen A. Schwarzman Building at the corner of Forty-Second Street and Fifth Avenue. One March day, I had tired of writing and emerged from the library around three o'clock, in plenty of time to beat the rush-hour traffic back to my home in Spanish Harlem. As I joined the queue for the bus, I stole a peek at my BlackBerry, where sad news awaited: Johnnie Cochran had died.

Almost immediately, a seventyish African American man joined the line behind me. He was thin, looked a bit frail, and wore an immaculate but slightly faded trench coat, a black fedora, and an expression of care and pain. As the bus arrived, he slowly mounted the steps behind me, sighing lightly. As I twisted around to make sure he was all right, he looked up and asked, "You heard about Johnnie Cochran?"

"Yes," I said as gently as I could. "It's a great loss." The man shook his head sadly and continued his ascent. Another man murmured something to him, and though I didn't catch the words, I understood the tone. This sadness was felt by many people, some of whom didn't care a fig about the exoneration Cochran had won for O. J. Simpson in 1995. African American men are so often demonized that it is hard to realize the extent to which they bear the brunt of bias in the workplace and in the health and justice systems, with few protectors. For many, Cochran had been a powerful legal champion who symbolized the possibility of long-delayed justice. But now he was gone.

As we took our seats, the man in the trench coat spoke again, more audibly, with an air of satisfaction. "You know, I'm from Anniston. Anniston, *Alabama*. They thought they could poison our homes and children, but Johnnie Cochran made them stop.

He made them stop, and he made them *pay*." The other nodded glumly, then they fell again into silence.

I'd all but forgotten his words when I heard the name again, this time at Cochran's funeral. In his eulogy, trial lawyer Theodore V. Wells capped a list of Cochran's triumphs for the downtrodden with "And today they've got that health center down in Anniston because of Johnnie Cochran."

I had to know more, and a little research quickly revealed that right after securing O. J. Simpson's not-guilty verdict Cochran had been approached by Anniston activist David Baker, executive director of the activist group Community Against Pollution in Anniston. Baker had explained that sickness was rampant among Anniston's inhabitants, thanks to extensive pollution caused by the Monsanto Company, the Olin Corporation, and even the U.S. Army. Even so, the Environmental Protection Agency and the Justice Department had utterly failed to pursue the offenders, leaving the citizens of Anniston—including children—to suffer with brain damage, lowered intelligence, and behavioral problems. Cochran agreed to help them obtain compensation for their losses.

He kept his word, organized a class-action suit, and procured for the victims the largest settlement ever won in the United States, including $50 million for a health clinic to address their rampant cancer, liver disease, and cognition problems such as memory loss.

As I read transcripts related to the suit, I was struck by Cochran's exclamation, "There is always some study, and they'll study it to death, then thirty years later, you find out it's bad for you. . . . We know it's bad for us right now!"[8]

In language that everyone can understand, Cochran was expressing an important precept that we have long shunned in the United States, called the *precautionary principle.*

Poisoned Minds

Approximately 60,000 industrial chemicals commonly used in the United States have never been tested for their effects on humans. In our country, safety tests are undertaken only when a chemical is suspected to be harmful. But even then, definitive findings are elusive, and it sometimes takes years or even decades of expensive research for them to emerge. Meanwhile, the standard of proof demanded by the industries that use and disseminate these chemicals is sometimes so high that masses of people suffer their effects in the time it takes to sufficiently prove their harmfulness. In the case of lead alone, the Environmental Defense Fund has noted that thousands of children were poisoned (at a cost of $50 billion to the nation) while we awaited "sufficient" proof to take action.[9]

As Cochran suggested, there is a better way. The European Union, for instance, requires human safety tests *before* any new industrial chemical is unleashed into the ground, atmosphere, and neighboring communities. It subscribes to the precautionary principle. In plain English, it is "better safe than sorry."

Because we have ignored this precept, lead poisoning has cost our country a staggering $50 billion. But it also has cost our nation something far more precious: 23 million lost IQ points *every year.*[10] And this frightful cost is the subject of this book, which discusses a specific, devastating, and all but unregarded injury that environmental poisoning inflicts on communities of color—the loss of intelligence and the malignant flowering of behavioral problems that destroy lives and human potential as effectively as cancer and lung disease but with far fewer alarms raised. IQ tests measure this loss of cognition, imperfectly to be sure, but in a useful manner.

The catalog of industrial chemicals known to erode intelligence is extraordinarily long, and many others are suspected of changing how well we think. In the chapters of this book, I reveal them. I delve into the deleterious effects of exposure to the witches' brew of chemicals—including PCBs, BPA, phthalates, volatile hydrocarbons, and more—that afflict adults, children, and the unborn within the nation's communities of color. I will discuss the cognitive costs of brain-hobbling microbes and, of course, exposure to heavy metals such as mercury and lead, which is blighting lives in the ongoing Flint water crisis.

Other common intelligence-lowering exposures include DDT and other pesticides that poison the soil, fish, and gardens of Triana, Alabama; the arsenic that turned neighboring Anniston into a postapocalyptic wasteland; the radioactivity spilling from Nanáʼáztiin uranium mines in New Mexico; the PCBs of Afton, North Carolina; and even the air pollution of Spanish Harlem.[11] These are just a few of the two hundred types of chemical exposure that have been shown to reduce intelligence and brain function by the work of scientists such as Philip Landrigan, Bruce Lamphear, and Philippe Grandjean, who dubs such intelligence-eroding chemicals "brain drainers."[12]

Medical journals offer evidence that many of the 140,000 untested industrial chemicals in worldwide use impede intelligence. As we will see, these chemicals are far more likely to find their way into African American, Hispanic, and Native American communities—affecting their water, land, and even schools—than into white communities. "Fence line" communities that abut toxin-belching industries, secret or open toxic dump sites, Superfund waste sites, and emission-belching diesel plants are preferentially located in communities of color, both poor and middle class. Robert Bullard, founder of the EPA Office of Environmental Equity, points out that the Moton School in New Orleans, for example, is built on top of a landfill, which leaches dangerous chemicals into the school's water supply, and similar hazards are rife in the nation's other ethnic communities: "In South Camden, N.J., schools and playgrounds are located on the waterfront alongside heavy industries of many kinds. Almost two-thirds of the children in that neighborhood have asthma. In West Harlem, the North River Water Treatment Plant covers eight blocks near a school. On the South Side of Chicago, it's the same kind of thing."[13]

The very young brain is not just that of a miniature adult. Fetal and infant brains are exquisitely sensitive to environmental poisons, especially ill-timed ones. This means that dismissing a tiny exposure as "harmless" because it is the equivalent of a drop of poison in 118 bathtubs full of water is naive. Such a vanishingly small dose is indeed enough to trigger lifelong disability in a fetus or infant, and in Chapter 4, I'll explain why.

Pathogens diminish intelligence, too, either by hampering the proper development of the brain or by directly assaulting the adult nervous system, or both. In Chapter 5, I'll show how and why these pathogens, from HIV to trichomoniasis, also preferentially afflict communities of color.

Children play against a background of the industrial effluents that dominate their neighborhood, one of the nation's many fence-line communities that are home to people of color. (AP Photo/Steve Helber)

In short, African Americans and other marginalized Americans of color are preferentially affected by chemicals known or strongly suspected to lower intelligence because they are far more likely to live in "sacrifice zones"—communities assaulted by environmental poisons and hazards. Addressing this chemical Armageddon will go a long way toward eliminating the "racial" IQ gap—without changing anyone's genetics. There are solutions we can work toward as individuals, communities, and as a nation, and I will note these as well.

It is important to realize that this book documents and discusses the greatly disproportionate impact of environmental hazards on racial minority groups, though it is not a burden borne solely by them. The recent finding that 90 percent of all Americans harbor pesticides or their byproducts in their bodies speaks to the universality of U.S. environmental exposures. But the reality is that race greatly exacerbates exposure and damages health—including both intellectual and mental health—in a dramatic manner. The intensity, breadth, duration, and harms of such exposures are greatly magnified in communities of color, although the news media and even scientific discourse have sometimes veiled this fact.

There are a few topics that I address in lesser depth than I originally expected to either because they go beyond the book's purview or because insufficient data limits an extensive discussion. Poverty, for example, is a potent thief of health and intellect, whether found alone or coupled with pollution. But although poverty likely potentiates disproportionate exposure to toxic environments, teasing out its influence independent of poisonous exposures requires research that has yet to be concluded. So instead of discussing its independent impact at length, I've pursued the most meaningful evidence-based discussion in this book's context: the relative effects of poverty and race in determining who is exposed to brain-eroding substances and agents.[14]

This Introduction and various later chapters address the cognitive harms of environmental agents on "poor whites" to the extent possible based on the available data. However, my ability to do so at length has been limited by the fact that publications on this group tend to focus on detrimental physical effects rather than cognitive ones. I was similarly disappointed by the dearth of data on farmworkers' exposures, even by the government agencies responsible for protecting their health. The Latino laborers and, on the east coast, African American migrant workers suffer greater proximate exposure than do most other Americans, but this cannot be quantified and analyzed without sufficient data.

Although this book discusses the effects of biological and psychological stress and cognition, it does not devote a lot of space to stress as an independent factor in cognition, which lies beyond the purview of a volume on poisoning wrought by the physical environment. Chapter 7, however, does explore recent research on the direct stress-mediating effects of the natural environment on mental health and cognition.

A Terrible Thing to Waste explains that establishing convincing proof, which is often presented as a purely scientific question, is very often an economic and political strategy deployed by industry and/or government to evade responsibility. Achieving scientific certitude is difficult given the plethora of variables and our often insufficient command of causal interrelationships.

But if protecting lives and health is what motivates us, perfect certitude may not be the gold standard, and we will not blindly dismiss strong evidence of untested chemicals as "mere correlation." When many studies of risk factors in the United States and abroad point to the same chemical culprits, this lends power to correlations and should be taken into account when deciding whether to approve their use.

IQ Antidote

If environmental poisoning is an important factor driving the IQ gap and what most scientists think it represents—some damping of intellectual performance—the news is not all bad. At the very least, this means that the gap can in fact be closed.

Nearly a hundred years ago, we learned that iodine supplementation can enhance IQ, yet we have failed to implement our knowledge, and so it remains the largest source of mental retardation. We cannot afford to make this mistake with environmental pollution. We can end human exposure to these poisons, and we must do so—for the sake of people of color, and for the sake of our nation, because we cannot afford to throw genius away. We must give the political polemics a rest and turn our energy from debating racialized genetic slander to applying what we know in order to level the playing field. We must topple the barriers to optimal intelligence for all Americans.

For more than a century, we've been assured that nothing can be done to restore the intelligence of groups who manifest evidence—such as lower IQ scores—that they are unprepared to compete in our fast-paced Western world, where the mastery of specialized skills such as reading and mathematical analysis has come to determine who will be a success and who will be resented as a burden on society.

But something *can* be done. Intervention is not futile. Once we recognize the large extent to which lower cognition is laid to addressable medical factors, it is easy to see that IQ loss can be corrected by reducing exposure to environmental toxins, and that our nation would reap huge rewards for doing so. Intelligence is a product of environment and experience that is forged, not inherited; it is malleable, not fixed.

By eliminating pervasive lead, mercury, hydrocarbons, industrial chemicals, prenatal exposures to alcohols, and even exotic pathogens like "red tide" algae poisonings, worm infestation, and trichinosis, we can save the assailed brains of untold people of color. As a bonus, by using techniques ranging from heavy metal and toxin abatement to educational enrichment to laws against industrial pollutants, we'll increase the health of our nation as a whole, and likely also help address the dangerous effects of dire poverty.

U.S. government action will also be necessary. A federal mandate could remedy the disproportionate racial exposures that hobble cognition. Closing or updating the last four mercury-based chlor-alkali factories in America, for instance, would save an estimated $24 million in economic productivity that is currently lost due to the human effects of mercury pollution.[15]

Unfortunately, the Trump EPA is moving in the opposite direction. In the waning days of 2018, it took aim at the Mercury and Air Toxics Standards (MATS), which under the Obama administration had funded the "$18 billion clean-up of mercury and other toxins from the smokestacks of coal-fired power plants." After a required sixty-day comment period, the EPA seeks to rescind the act, which it deems not "necessary," although it and similar efforts have reduced mercury from coal-fired plants by a heartening 85 percent over a decade.[16]

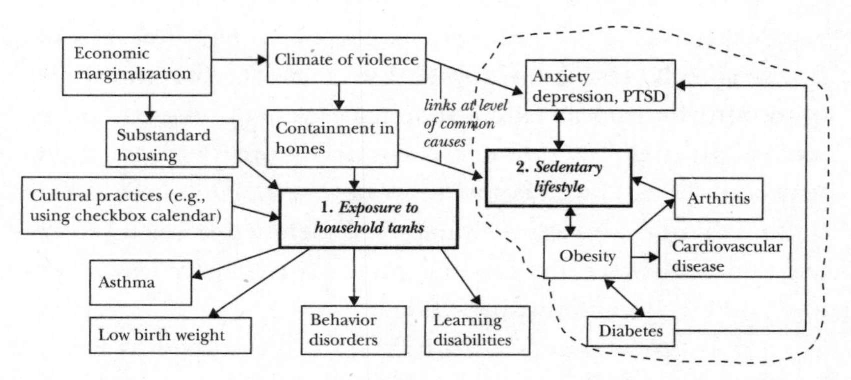

Environmental toxicity greatly erodes intelligence and affects behavior in concert with many other factors, as this chart illustrates.

The National Debt

Somehow the misperception has arisen that lowered IQ in ethnic enclaves affects only the people who live in them. However, as developmental toxicologist Bruce Lanphear has shown, a small five-point decrement in collective IQ not only drags down the average intellect of a nation, it slashes in half the number of people who fall into the "gifted" category while swelling the population of the mentally retarded or intellectually disabled by an additional *3.4 million.*[17] The IQ gap's persistence in the United States is part of the reason our intellectual rankings lag behind other Western societies with fewer resources. We rank on par with countries like Estonia and Poland, while top scores go to Japan, Australia, and Scandinavian countries.

Some question how critical IQ is. We've long known that IQ measurements, in the United States and around the world, are dramatically biased. We also know that it is not possible to administer the test in a manner that gives meaningful comparisons across a wide variety of cultures. Beyond this, the meaning of "intelligence" varies from culture to culture, it is multifactorial, and IQ tests provide an admittedly limited and biased measure of achievement, not the oft-touted innate ability.

In the context of this book, which focuses on the United States, it is exactly this testing of achievement that imparts some meaning to IQ tests. Christopher Eppig, Ph.D., director of programming for the Chicago Council on Science and Technology, muses, "In the United States, the ability to use technology, to read and to use numbers is important, and IQ tests measure these well. In other, less technological societies, lower IQ scores may reflect the fact that reading and calculation are less important and less frequently practiced."

"I don't know how predictive IQ is of your success as a good hunter-gatherer or farmer," notes Brink Lindsey, author of *Human*

Capitalism: How Economic Growth Has Made Us Smarter—and More Unequal. "But it predicts whether you'll be a good office worker."[18]

Our nation has relatively few job openings for farmers. In today's technological society, the species of intelligence measured by IQ is what's deemed most germane to success.

Although IQ scores are not a consistently accurate measure of intelligence, IQ is too important to ignore or to wish away. For Americans, IQ, usually measured by the Stanford-Binet and Wechsler scales, although there are many variants, has proven a predictor of success in school, social settings, work achievement, and lifetime earnings, while emotional intelligence scores, for instance, are far less meaningfully linked to success.[19] The median hourly wage of workers scoring on the highest level in related literacy tests, for example, is 60 percent higher than that of workers scoring at the lowest level. Those with low literacy skills are more than twice as likely to be unemployed.

It's true that the association between IQ and academic and material success is less strong for marginalized minority-group members because racial bias strongly drives success and failure as well. White men with high school educations earn more than black men who graduate from college, for example. But for minority-group members, who are more likely than other Americans to be poor and underemployed, lower levels of achievement or income mask an especially dire situation.

Globally speaking, the average American of any race or ethnicity is in trouble. In 2013, the Program for the International Assessment of Adult Competencies found that American adults scored well below the international average in math, reading, and problem-solving abilities using technology: 10 percent cannot use a computer mouse.[20]

Jacob F. Kirkegaard, a senior fellow at the Peterson Institute for International Economics, summed this up: "There is a race between man and machine here. The question here is always:

'Are you a worker for whom technology makes it possible to do a better job or are you a worker that the technology can replace?' "[21]

Recent research has confirmed that IQ predicts the growth and future success of nations, too. A 2014 Malaysian University study strongly suggests that people with the highest IQs have the most powerful influence on economic development, even more than professionals engaged in research and development activities. But our unimpressive scores mean that Americans are less able than most people in other developed countries to solve problems in a technological environment.

This isn't going to change anytime soon. The United States will continue to face difficulty catching up with China and India, our nation's current economic competitors, in part because state education funding has been slashed in recent years.

The Plea of Impossibility

Yet, as some of us seek answers, futility is the common and disheartening refrain of the hereditarians, whose opinions have most fundamentally shaped the IQ debates. They claim that intelligence is innate and immutable and insist that nothing—not better health, not education, not reducing poverty, and presumably not removing the assortment of poisons assailing the bodies and brains of children—can be done to raise the IQs of ethnic groups.

Nearly all of them specifically argue that investing in solutions to the IQ gap is a waste of money and resources, ignoring the successes of solutions like complete lead abatement (which has been demonstrated by longitudinal studies to preserve children's IQ) or pre-K enrichment programs such as Head Start (which are associated with bounds in academic performance).

These unsupported claims of futility serve their proponents' various political agendas well, such as stemming Hispanics' immigration rather that enriching their education so it meets national standards, or explaining African Americans' second-class status as a function of their own innate limitations rather than as a function of systemic racism.

But this fictitious futility does not serve the needs of the lead-poisoned boy who cannot learn to read, the girl whose desultory education has left her unprepared for the workplace, or the desperately poor rural family whose hookworms and privation have left their children with subtle brain derangement that may never be diagnosed, as I discuss in Chapter 5. Assuming that "nothing can be done" is a self-fulfilling prophecy: nothing *will* be done by a nation that adopts this stance. Hereditarians well understand that clinging to the notion of futility absolves them from the responsibility to act.

British philosopher Jeremy Bentham said it best: "The plea of impossibility offers itself at every step, in justification of injustice in all its forms."

Embracing futility sabotages the welfare of our nation on all fronts. *A Terrible Thing to Waste: Environmental Racism and Its Assault on the American Mind* shifts the focus from political sniping to what science tells us about real, correctable threats to intelligence and about strategies to maximize the intelligence of people of color and of the nation as a whole. A wealth of empirical evidence points the way, but both camps have given it short shrift. It is imperative to be guided by data rather than by entrenched political agendas or the trading of ideological insults.

We must also learn from past mistakes, such as the penny-wise and pound-foolish partial abatement of lead, mercury, and other intellect killers, in order to fully invest in our children and our nation's future. Despite the tensions that have driven our perception of this problem, we will rise or fall together, not as separate ethnic groups.

Nations with higher average IQs have stronger economies

and greater wealth. But money isn't everything: A 2014 report into the relationship between IQ and national progress noted, "Empirical studies have found that intellectual people, namely those that have a high IQ, contribute more to socioeconomic development in a society as compared to the average ability citizens."[22]

We ignore this fact at our peril.

PART I

Color-Coded Intelligence?

CHAPTER 1

The Prism of Race: How Politics Shroud the Truth about Our Nation's IQ

I am, somehow, less interested in the weight and convolutions of Einstein's brain than in the near certainty that people of equal talent have lived and died in cotton fields and sweatshops.

— Stephen Jay Gould[1]

As she stood in the doorway of the class surveying a sea of small, mostly dark, faces, Cathy* suddenly realized that she was clutching her handbag tightly, too tightly. She forced herself to relax and offer a smile she did not feel. "I don't understand. Why is Edgar in a special education class?"

Cathy knew her son was bright. He loved school, read well above his third-grade level, and, in New Jersey, he had always hovered near the top of his class. She had ascribed his recent moodiness and reluctance to attend school to difficulty adjusting to their move to Fort Lauderdale, but when she prodded him, he blurted, "They have me in the class for slow kids, like I'm a dummy. Some of those kids can't even read!"

---

* All of the names and some of the details in this composite vignette have been changed to preserve privacy.

Incredulous, Cathy dialed Teri Marche, Edgar's teacher, and was appalled to hear that he had indeed been placed in special education. "Why did no one tell me?" she demanded. "We need to talk."

On her ensuing tour of the school, Cathy noted that nearly all the children in the gifted program were white, and asked, "On what test scores are these placements based?"

Teri explained that each instructor's assessment of the student, not standardized tests, determined placement. Cathy insisted, "My child is very intelligent. Test him."

"Please, we can optimize Edgar's training in these focused classes and he'll perform better in the end. We must be realistic: the IQ gap is a fact, and that, unfortunately, explains the racial disparity we have here."

Training? Now Cathy was really angry, and she whirled around to face Teri. "You don't *know* my son's IQ. I promise you that if he is not tested I will pursue legal recourse." She turned to leave. "Thank you for your time."

The next week, Edgar was tested and reassigned to a gifted class. In the following year, 2005, the Broward County Department of Education began administering standardized tests to all second graders.

As a result, the number of gifted African American children in the district soared 80 percent, and that of gifted Hispanic children skyrocketed 130 percent. Shortly thereafter, the policy shift resulted in a tripling of both black and Hispanic "gifted" students.[2] None of the students' IQs—or genes—had changed.

Mind the Gap

The average 15-point black-white gap in measured IQ—and the sort of rampant assumptions that Cathy faced—are more than fodder for political debates about inherent racial intelligence. The widespread (and inaccurate) belief that IQ limits are innate and permanent directly affects the trajectories of lives like Edgar's, often imposing a low ceiling on an individual's opportunities from an early age, despite our knowledge of environmental risks to IQ, such as iodine deficiency. The fact that iodine deficiency was responsible for a 15-point IQ gap between residents of different geographic areas should have impressed upon us the environment's power to create the same 15-point IQ gap between U.S. whites and African Americans, especially considering the extent to which this country remains geographically segregated along racial lines.

We also should have noted that addressing a damaging environmental factor is an effective tool that allows us to close the gap. Iodine deficiency is far from the only environmental condition known to lower IQ, and it is very far from the most potent. Heavy metals such as lead, mercury, and arsenic, and nearly 150,000 untested industrial chemicals including pesticides, phthalates, PCBs, halogenated hydrocarbons, alcohol, and more have been shown to depress IQ, stymie intelligence, and distort human behavior. Moreover, copious scientific data documents how people of color suffer the effects of these "brain poisons" in wildly disproportionate numbers. Industrial waste dumps, workplace exposures, Superfund sites, and bus depots haunt their neighborhoods. Scientists are beginning to quantify these effects in a manner that supports the connection between IQ and environmental exposures in communities of color.

Yet we have largely failed to appreciate the well-documented connection between environmental exposures and lower intelligence and IQ. Certainly we have failed to act upon it. Acknowledging the potent effect of environmental poisons on intelligence would contradict the hereditarian theories of innate, permanent inferiority as expressed by IQ scores, theories that are uncritically accepted by many. The hereditarian view of IQ dominates the national perception of IQ as a measure of human potential. As I mention in the Introduction, hereditarianism's racially loaded position makes for grabby headlines, and those who oppose it spend a great deal of time in point-by-point refutations of arguments and insults, so that instead of promulgating a more accurate view of intelligence and IQ we get a shrill, reductionist national conversation.

This myopia robs us of a potent tool—the reduction and removal of human exposures to environmental poisons that cause damage far beyond cancers, lung disease, and other serious physical illnesses. They also damage the brain and lower intelligence, creating not only individual tragedies but a needless collective loss of intelligence.

For a better understanding of what IQ is and isn't, let's go back to the beginning for a brief look at the nature of intelligence and IQ.

Intelligence seems necessary for material and social success, and is central to our idea of success in general: personal freedom, good grades, holding a meaningful, profitable job, providing for our families, maintaining a rewarding social network, financial security, and generally navigating the complexities of modern life in the West. However, if you're not sure what, exactly, intelligence is, don't worry: You're in good company. When asked to look up from their tools and metrics to offer a definition, scholars who've devoted careers to refining our understanding of this mercurial attribute—or collection of attributes—struggle to define it.

"That's a good question," responded one researcher I interviewed, pausing briefly before adding, "A very good question," and falling into silence.

Christopher Eppig responded, "The best definition that I've heard of intelligence is that intelligence is an emergent property of mental functioning. You can compare it to attractiveness, which is not based on one single trait like the distance between someone's pupils. It emerges from thousands of traits upon a person's body and mind and it's not even the sum of lots of things, but it's the interaction of these things that gives you an emergent property."[3]

"A good definition of intelligence? That's a great question," said Brink Lindsey. "I don't know. There's really no definition except 'performing well on IQ tests,' although there is the speed of learning and retention and how fast one processes information and how well one retains it."[4]

In 1969 Arthur Jensen threw down a modern hereditarian gauntlet when he alleged the genetic intellectual inferiority of African Americans to whites. In an essay for the *Harvard Educational Review* entitled "How Much Can We Boost IQ and Scholastic Achievement?," Jensen opined that IQ scores indicate unchangeable, inborn, genetically inherited intelligence for all groups of people. Differences in IQ scores between races, he claimed, prove group differences in fixed native intelligence.[5]

But when asked to define intelligence, Jensen sidestepped the question, intoning, "Intelligence, like electricity, is easier to measure than to define."[6] Most high school physics students could refute this nonsensical aphorism by explaining that electricity is a form of energy resulting from charged particles, either statically, as an accumulation of charge, or dynamically, as a current.

IQ Rankings, by the Nation

Jensen's inability to define intelligence didn't stop him from joining his forebears in ranking some Asians and Europeans near the top and the diverse denizens of Africa—from Nigerian urbanites to Tanzanian scientists to Kalahari bushmen to Congo "pygmies" to Ethiopian Jews—near the bottom of an IQ gradient that persists today.

Researchers write of intelligence, but they measure IQ, and these are not synonymous. "There is no direct test of general mental ability. What IQ tests measure is the display of particular cognitive skills such as vocabulary and reading comprehension," explains Christopher Eppig. "Any conclusions about general mental ability are inferences drawn from the test-taker's mastery of those various skills."

In 2002, when *IQ and the Wealth of Nations,* by Richard Lynn and Tatu Vanhanen, ranked nations by their average IQ scores, most of Europe and the more affluent portions of Asia hovered a bit above 100 points. But the news for Africa, Hispanic nations, and South America was grimmer. Average IQs for most African states fell below 70 (except for South Africa, which came in at 72), although Ethiopia's score of 63 was later revised to 71. Equatorial Guinea defined the nadir at 59. The authors suggested that these nations' low average IQs denoted low intelligence among their citizens, which in turn explained the poverty and scant economic development in sub-Saharan countries. Low intelligence, indeed. By most measures, an IQ below 70 denotes mental retardation; thus Lynn and Vanhanen claim that nearly all of Africa is mentally retarded.

Critics pointed out that their conclusion was unsupported by the data. Their methodology lacked rigor, was illogical,[7] and their IQ calculations used very small, non-representative populations and were often derived from older papers and assessments of dubious credibility.[8]

But even if their methodology had been sound, their conclusions would still have been unreliable. As Brink Lindsey points out, the very concept of comparative IQ scores is flawed.

> [IQ] scores are only a good indicator of relative intellectual ability for people who have been exposed to equivalent opportunities for developing those skills—and who actually have the motivation to try hard on the test. IQ tests are good measures of innate intelligence—if all other factors are held steady. But if IQ tests are being used to compare individuals of wildly different backgrounds, then the variable of innate

intelligence is not being tested in isolation. Instead, the scores will reflect some impossible-to-sort-out combination of ability and differences in opportunities and motivations.[9]

Girma Berhanu, a professor of education at the University of Gothenburg, agrees, adding, "Statistics can lead us to accurate conclusions only by using representative samples that are selected at random." In Ethiopia, however, the average IQ score was calculated based on a sample of a mere 134 children in a single orphanage in Jimma. As Berhanu wrote, "An orphanage in Jimma in 1989 was an extraordinary and traumatic experience for children who were victims of famine, resettlement, and on a massive scale. The experience of orphaned children who survived harrowing experiences of death and starvation cannot be seen as a representative sample for IQ testing."[10]

Thus, Lynn and Vanhanen's rankings are nonsensical because the developing-world sites used in establishing IQs have so very little in common with Western and affluent Asian nations. In many such countries, compulsory education and literacy are relatively rare, and not all the tested subjects were fluent in English or in the language in which the test was administered.

As Berhanu points out, social conditions in very poor and unstable countries can militate against success in test-taking. The incentive to perform well on such lengthy, grueling tests is strong for Americans, who know that their future success in academia and industry hinges on a high score. This motivation doesn't exist among test subjects in developing nations who are almost certainly correct in seeing no connection between a high score and their lot in life.

Moreover, health status stands between the people of the global South and a normal IQ score. Poor nutrition alone depresses brain development and intellectual functioning. So does exposure to neurotoxic agents, which is ubiquitous in countries where industrial emissions are rampant and all but uncontrolled. This means that even had testing practices in these countries been logical, meticulous, and fair, environmental factors alone would still cause a disparity in IQ scores.

In this respect, the denizens of low-IQ nations have a lot in common with residents of lower-IQ ethnic enclaves of the United States. They suffer exposure to the same sort of brain-damaging chemicals as do members of U.S. racial minority groups and with the same potent but underappreciated effect on their cognition and intelligence as is apparent in their depressed IQ scores.

The various IQ tests used in Lynn's book and still in use elsewhere have something else in common: They measure abilities that are important to success in Western culture and that are more commonly practiced in relatively affluent Western cultures than in other nations and cultures. The tests, for example, focus heavily on reading skills, which are not important in every culture. Developing nations that ranked low in IQ often lacked compulsory schooling, and many of those tested were rural agrarians who had rarely if ever held a pencil.

Some dismiss such rankings as not only scientifically flawed but racist, or at least ethnocentric, especially because many of the hereditarians who espouse them are funded and supported by politically active groups with eugenicist or anti-immigrant leanings, such as the Pioneer Fund. (It's worth noting that Asians, not Europeans, often occupy the very apex so that these rankings serve to denigrate dark-skinned people, not to elevate the palest people.) Yet "racist" can be an unhelpful assessment. Surveys show that various groups define the term quite differently, so it often impedes rather than clarifies communication. Moreover, even if such IQ assessments are motivated by racial disdain, this doesn't prove them wrong. IQ is a crude, inaccurate, and biased proxy for intelligence, but most scholars agree that it measures *something* to do with the capacity to learn, at least in the affluent West.

The error of assuming that IQ scores indicate inborn lower intellectual capacity dictated by race is a product of bias, intentional or otherwise, that inaccurately depresses some groups' IQs.[11] However, this belief is also due to misunderstandings about what IQ can and cannot measure.

If, for the sake of argument, we assumed that the global rankings are more accurate and valid than they are, then there would be a reason that the IQ of the average Ethiopian is lower than that of the average U.S. resident. And knowing what this reason is would be a necessary first step toward a corrective.

Hereditarians say the reason is that inferior IQ means that the intelligence of racial groups such as African Americans and Hispanics is

- innate,
- primarily genetic and so impervious to prevention, and
- immutable, and cannot be ameliorated through better education, or better physical and mental health.

Perhaps these assumptions, not the findings of lower relative IQs, are the most damning tenets of hereditarian dogma. Perhaps it is not the fictitious assessment of lower IQs that is most malignant, but rather the unsupported assumptions about the nature of IQ and intelligence that invest the IQs with dire meaning.

The Rising Tide

Fortunately, the hereditarians are wrong: The facts belie all their beliefs.

IQ is not innate. Studies that compare the IQs of mixed-race children reveal the determinant role of their various environments. In Germany, children with one parent of African descent and one of European descent who were raised in white households (such as the offspring of African American or African soldiers whose fathers return to the States and whose mothers subsequently marry white Germans) scored higher on IQ tests than mixed-race children from similar backgrounds who were raised in black households. The same thing happened when comparing mixed-race children in the United States.[12]

IQ is not "genetic." Most experts consider intelligence to be multifactorial, and no genes have been proven to determine IQ.[13] Furthermore, although we know that genetics are an important factor in most aspects of human health, the early experience of the brain and nervous system is far more important when it comes to intelligence. Pertinent studies include a German research study in which the overall average IQ of biracial boys with a white father was found to be 96, while that of such boys with a black father was found to be 96.5: virtually identical, even though the latter group's boys were subjected to racial prejudice.[14] A U.S. study compared black and biracial children adopted by middle-class black or white parents with similar results.

Genetic potential is just that—potential—and whatever potential a given fetus has, exposures and deprivation can ensure that his or her future IQ will be lower than average, sometimes devastatingly so. Whether a child in utero is exposed to alcohol during a key moment of development, assaulted by PCBs, arsenic, or mercury, or assailed by a pathogen associated with the loss of cognition or memory, it is the brain's experience, not genetics, that will dictate the child's intelligence and IQ.

IQ is not impervious to change. We have been watching average IQs change within populations for decades in the West as well as in some of those "irremediably low-IQ" nations.

Consider Kenya. IQ tests were administered twice in certain regions of Kenya, once in 1984 and again in 1998. The later test reported an average IQ at least 14 points higher than the first. This increase occurred far too swiftly to have been driven by genetics. However, this spike in average IQ coincided with a period of dramatic improvement in health status in the regions, including lowered rates of infectious disease. It is highly unlikely that these improvements were unrelated.

In fact, this IQ increase is a variant of the "Flynn effect," named for James R. Flynn, whose 2012 book, *Are We Getting Smarter? Rising IQ in the Twenty-First Century,* documented an average IQ rise of three points per decade between 1932 and 1978 in the United States and some other Western nations. Subsequent analyses have confirmed this, such as a 2014 meta-analysis in the *Psychological Bulletin,* which calculated the same IQ gains between 1972 and 2006.[15] This suggests that the scores reflect not innate intelligence but acquired skills.

So IQ is indeed malleable—we just need to determine what causes it to change. If we do that, we will be empowered to craft strategies and tools for raising it within marginalized ethnic minority groups of African Americans, Hispanics, and Native Americans, just as we raised the low IQs within pockets of America by using iodine supplementation.

Necessary Bias

If IQ measures acquired skills (rather than innate ability), if it is affected most strongly by environmental forces (not transmitted by genetic ones), and if it depends upon a shared culture and strength of motivation to tell us anything meaningful about ability, why do some cling to notions that a racial difference in IQ means an innate, unchanging difference in African American intellectual potential?

To understand this, it is important to realize that the belief in the innate, irreversible, intellectual inferiority of people of color predated any objective tests or measurements of their abilities. As I explain in *Medical Apartheid: The Dark History of Medical*

Experimentation on Black Americans from Colonial Times to the Present, the assertion that African Americans and some other people of color were inferior and perhaps not even human was a necessary myth for justifying enslavement, and was supported by the nineteenth century's most prominent U.S. scientists, the American School of Ethnology. Intellectual inferiority was a key tenet of that inferiority, repeated often in the scientific and popular literature.[16]

Nineteenth-century scientists portrayed the enslaved African American as inherently debased and permanently so: no amount of training, education, or good treatment could make him the equal of a white man. Most of these scientists were polygenists, who thought that African Americans and whites belonged to different species. They held that black Americans were physically inferior, dishonest, malingering, hypersexual, and indolent. This inferiority was documented in entire catalogs of black flaws that filled medical journals and textbooks.[17]

In 1839, Samuel George Morton published *Crania Americana,* a book written to demonstrate how human skull measurements, or craniometry, revealed a hierarchy of racial types. Morton determined that Caucasians had the largest skulls, and therefore the largest brains, and blacks the smallest. His tests were the forerunner of phrenology, which sought to determine character and intelligence by interpreting the shape of the skull.[18]

Nine years later, Louisiana's Samuel A. Cartwright, M.D., suggested that blacks' physical and mental defects made it impossible for them to survive without white supervision and care, alleging that the crania of blacks were 10 percent smaller than those of whites, preventing full development of the brain and causing a stunting of the intellect.

French scientist Louis-Pierre Gratiolet added that in the Negro "the cranium closes on the brain like a prison. It is no longer a temple divine, to use the expression of Malpighi, but a sort of helmet for resisting heavy blows."[19]

IQ is measured by a myriad of examinations that purport to evaluate and score a person's intelligence—partial or entire. Haunted by race bias, class bias, and a plethora of common misapplications, the IQ test is a widely applied but very imperfect yardstick, useful in measuring intelligence, but only within a limited context.

And how could it be otherwise? A pervasive American belief in the intellectual inferiority of people of African descent and of formerly colonized peoples predates any tests formulated to document it. These include people we would now group under the labels of Hispanic Americans, as well as some Asian Americans and Pacific Islanders. From the beginning, racial measurements of intelligence have been plagued by rigged research and other misapplications of science designed to enforce racial bias, as Stephen Jay Gould documented extensively in *The Mismeasure of Man* and as Robert V. Guthrie reveals in *Even the Rat Was White: A Historical View of Psychology*. Continued belief in the inferiority of people of color has survived even the grossest logical failures and embarrassing revelations of racial bias.[20]

Mental Fitness by Numbers

National data used to dictate policy have also been manipulated in bad faith, as when the 1840 U.S. census employed falsified data to promulgate a pro-enslavement message of racial mental inferiority. Census data purported to show that freedom was unhealthy for the limited minds of African Americans and caused freedmen to suffer *eleven* times the mental disease of their enslaved peers.

It took James McCune Smith, M.D., a Glasgow-trained African American physician and statistical scientist, to refute the scientific racists on their own ground, and he found it necessary to do so frequently. His 1837 lecture exposing the scientific fallacy of phrenology offered scathing criticisms of the logical sins inherent in imputing character and intelligence from physiology, and his 1844 report to the U.S. Senate entitled "The Memorial of 1844 to the U.S. Senate" exposed the intentional bias and manipulation used to impugn African American intelligence in the 1840 census data.[21]

Guthrie reveals many instances of such racial bias in test taking. Among them, the administration of "identical" intelligence tests to armed forces recruits in the 1920s. Whites and blacks were given the same test, but middle- and upper-class whites took a paper-and-pencil "alpha" test whereas blacks and some poor whites were assumed illiterate, so the test questions were "given by demonstration and pantomime"[22]—acted out—by proctors in the "beta" variant of the test.[23] Clarity was sacrificed because at least some abstract test questions were utterly unsuitable for miming.

Other Army tests utilized "picture completion" exercises that rated the soldiers' ability to draw the missing element in an image such as the balls in the hands of bowlers and the missing net in a depiction of a tennis game. For a sharecropper's son on his first foray outside the deep South these were likely to be unfamiliar objects. Among the black and lower-class participants, the test generated a great deal of confusion, and very low scores. W. E. B. Du Bois observed:

> For these tests were chosen 4730 Negroes *from Louisiana and Mississippi* and 28,052 white recruits *from Illinois* [emphasis original]. The result? Do you need to ask? M. R. Trabue, Director, Bureau of Educational Service, Columbia University, assures us that the intelligence of the average southern Negro is equal to that of a 9-year-old white boy and that we should arrange our educational program to make "waiters, porters, scavengers, and the like" of most Negroes![24]

In the early twentieth century, W. E. B. Du Bois criticized the Army intelligence tests, which supported belief in innate African American intellectual inferiority. (Public domain)

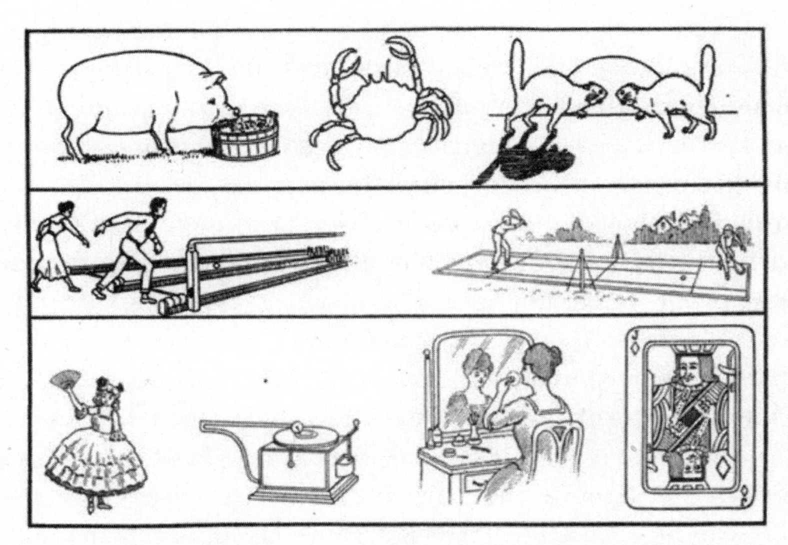

Picture completion exercises, part of the Army intelligence tests, often relied upon knowledge (such as the proper placement of a tennis net or powder puff) that would have been foreign to many poor black recruits, resulting in lower scores for them. (Courtesy of Arlene Shaner, New York Academy of Medicine)

In an often-mystifying display, the proctors "acted out" instructions for the early twentieth-century beta Army intelligence test instead of giving verbal directions. The beta test was ostensibly designed for the illiterate but was given indiscriminately to "negro" recruits. (Courtesy of Arlene Shaner, New York Academy of Medicine)

During the 1920s, eugenicists influenced the passage of the National Origins Act of 1924, which barred immigrants from Southern European countries as "dysgenic." This line of thinking persists to this day: within the past decade, hereditarian Jason Richwine sought to convince lawmakers to similarly restrict immigration from Hispanic nations partly on the basis of their supposedly lower intelligence. Richwine's 2009 Harvard doctoral dissertation, entitled "IQ and Immigration Policy," claimed that Hispanic immigrants had lower IQs than non-Hispanic whites, a disparity that he claimed persisted for generations.[25]

The belief that IQ scores are hereditary, that they measure general intelligence, that they are innate, and fixed, and that they can be compared across groups are assumptions that have been made for more than a century, without proof. Today, they are made in spite of proof to the contrary, as recent research has roundly dismantled most of these myths about IQ. Such myths should never have been embraced. In fact, they contradict the very research, goals, and findings of the man who invented the notion of IQ.

IQ Origins

Psychologist Alfred Binet, originator of the twentieth-century intelligence test known as the Stanford-Binet Intelligence Test, was frustrated in his initial attempts to create a bias-free intellectual assessment based on craniometry, the standard method at the time. He traveled from school to school measuring cranial sizes and attempting to correlate them with student performance, noting that teachers made the subjective assessments of which students were stellar or underperforming. But such determinations were validated by no objective measure.

Binet found only minuscule differences in performance relative to cranial size: mere millimeters separated the brightest from the dunce, and this was not useful in predicting intellectual performance.

In 1904 the French minister of public education asked Binet to find ways of identifying the specific cognitive problems affecting poorly performing schoolchildren so that ways of addressing and improving their individual learning issues could be devised. Binet did so, this time rejecting craniometry in favor of a tool he developed himself, specifying that it was to be used only to evaluate the learning disabled and low academic achievers. He assigned children brief, scored tasks of counting and reasoning and used the scores to determine what he called their mental age. By subtracting this mental age from the child's chronological age Binet arrived at a quotient, an individual measure of academic achievement. The intelligence quotient, or IQ, was born.[26]

From the beginning, Binet insisted that the test was valid only for measuring the weaknesses of pupils who were already performing inadequately, not for predicting who might perform inadequately. He always denied that it could be used for the latter purpose. In fact, writes University of Wisconsin–Madison professor of law and bioethics Pilar Ossario, "[Binet] rejected the notion that his test measured a person's inborn or fixed cognitive ability. He also declined to use his test to rank individuals according to cognitive ability."[27]

"What Binet feared most, about an IQ number," wrote Robert Anemone, a biological anthropology professor at the University of North Carolina at Greensboro, "was its negative uses in society. He thought that it could be used as an indelible label rather than a tool to identify the needs of the child.... Therefore, Binet declined to label IQ as inborn intelligence and refused to regard it as a device for ranking individuals based on the mental capacity."[28]

Of course, this is precisely how the IQ test is used today—as a device for ranking individuals' mental capacity and stamping them with an indelible label.

While a history of IQ testing is beyond the scope of this book, it is helpful to understand that later psychologists adopted Binet's IQ test for the very different purpose of assessing the intellectual capacity of anyone. And this was just the first of many misinterpretations of what IQ can and cannot tell us. Thus, the chief principles of Binet's IQ test were reversed as early psychologists translated his scale into a universal tool for testing all children. They further assumed this score to be fixed for a person's entire life, and hereditary—handed down from parents to child—all without evidence.

So do contemporary American hereditarians. Twin fallacies drove this misapplication of the IQ test. Reification, which bestowed an independent determinative and predictive reality and power on the test; and hereditarianism, the assumption that IQ is inherited, passed from parents to child.

These assumptions fit well into the pervasive preexisting racial mythologies that already held the intellectual inferiority of the "lower races" to be rigid and heritable. IQ became another tool to validate and reinforce their lower station in life.

Today many intelligence tests give a single intelligence quotient, held to be a measure of general intelligence. IQ tests purport to measure only *fluid* intelligence. Fluid intelligence is associated with the ability to solve problems that require no prior knowledge, such as abstract logic puzzles that one has never seen before and for which one has no context. *Crystallized* intelligence, on the other hand, is related to one's ability to use learned knowledge. IQ questions regarding vocabulary and analogies, which can more readily be answered by those who have read widely and know a lot of words, fall into this category.[29] So IQ tests actually measure both fluid and crystallized intelligence. As Brink Lindsey notes, "Fluid scores peak by the twenties but crystallized intelligence continues to improve until approximately age sixty."[30] Head Start and other pre-K enrichment programs boost both crystallized and fluid intelligence.

Other tests measure discrete aspects of intelligence, such as verbal abilities. The SAT, which measures crystallized knowledge such as literacy, mathematical and writing skills, and the ability to solve problems that are needed for academic success in college, is often used as a proxy for an intelligence test, even though the SAT does not purport to measure intellectual potential but rather acquired skills.

The hereditarian arguments now hinge upon an average 15-point gap between U.S. white and African American IQ scores. Because the intelligence IQ measures is genetically passed from parent to child, they argue, the lower IQ scores of African Americans must be genetic and thus cannot be improved by early or intensive education. Some people even assume the average gap between African American and white scores applies to all individuals, as Edgar's teacher, described earlier in this chapter, may have done when she cited the IQ gap as the reason for his baseless school placement.

Hereditarian claims are refuted in detail by contemporary works such as J. L. Kincheleo et al.'s *Measured Lies,* C. S. Fisher et al.'s *Inequality by Design,* M. K. Brown et al.'s *Whitewashing Race,* and R. E. Nisbett's *Intelligence and How to Get It.*[31] Invoking an average African American IQ score that is 15 points lower than that of whites errs in its assumption that the difference is innate. We've seen such assumptions of genetic mediation invoked before, as Stephen Jay Gould reminds us. Before World War II, the shorter stature of the average Japanese national was assumed to be a racial, genetically transmitted trait. But after Westernization and economic prosperity led to an enriched diet, the average height of the Japanese increased dramatically, and far too quickly to be ascribed to genetics.

Not only are the hereditarians' claims wrong, their arguments are illogical and reflect naivete about genetics. They often espouse futility and urge the defunding of educational enrichment programs such as Head Start because, they insist, nothing can change "genetically" lower IQs. But not only has no specific gene been identified that mediates the IQ differences they reference, these enrichment opponents are conveniently forgetting a key fact about genetic disorders and conditions: "Genetic" does not equal intractable; so genetic conditions are not by definition "without remedy." Even biological flaws due to demonstrably genetic causes can be remedied. For example, children born with the genetic disorder phenylketonuria (PKU) suffer profound mental retardation—unless they are diagnosed in time to withhold phenylalanine from their diet, in which case they suffer no retardation or ill effects at all. So even if IQ *were* genetically mediated, this would be no reason to abandon efforts to elevate it.

Gould also offers the example of nearsightedness, a visual defect that can be inherited and pose a serious bar to achievement if it is severe enough to prevent the affected person from learning, reading, functioning well on the job, and even negotiating the world safely. But eyeglasses and contact lenses erase the disability and close the performance gap between the nearsighted and the normally sighted.

Healthy, Wealthy, and Wise?

I submit that in today's communities of color IQ is largely compromised by disproportionate health assaults like a legion of chemical exposures, nutritional deficiencies and poisons, alcohol, drugs, and pathogens that cripple the nervous system. In *IQ and the Wealth of Nations,* Lynn and Vanhanen agree that intelligence is causally related to economic and health status, but they argue that high intelligence *creates* well-being, not the other way around. They also insist that beyond improving nutrition, very little can be done about poor health caused by low intelligence.[32] But as Girma Berhanu wrote in the journal *Educational Review,* "Vanhanen has got his argument backwards. It makes far more sense to argue that the populations of rich countries do better on IQ tests because they have access to better nutrition and education."[33]

In 2006 evolutionary psychologist Satoshi Kanazawa pushed the theory of reverse causation further by theorizing that the low intelligence among people of color in the developing world has created their poor health status. In "Mind the Gap...in Intelligence: Re-examining the Relationship between Inequality and Health," published in the *British Journal of Health Psychology,* he posited that distance from the African birthplace of man directly correlates to both IQ and to health status. "The farther away a nation is from sub-Saharan Africa, both latitudinally and longitudinally," he added a few years later, "the higher the average intelligence of the nation's population."[34]

According to Kanazawa, a reader in management at the

London School of Economics, Africa, and by extension the larger tropical and subtropical world, provided a gardenlike environment that was warm and nurturing with abundant food—familiar and easy to live off, he argues, so that success within it requires very little intellectual effort. But those early men and women who left the continent for the north and west encountered unfamiliar environments characterized by freezing winters and seasonally disappearing food—environments that Kanazawa says were far more difficult to survive. Evolutionary pressures selected for those clever enough to adapt, so these people became healthier, smarter Westerners, including scholars who now congratulate themselves on having picked the right forebears.[35]

But experts point out that Kanazawa's basic assumption is erroneous. Africa is no benign Garden of Eden with fruit hanging from every tree. Africans had to contend with apex predators, poisonous plants and animals, drought, oppressive rainy seasons, unfriendly neighbors, and, in the Southern and elevated parts of the continent, with snow and cold, just as in Europe. Not to mention a prodigious variety of infectious diseases. University of New Mexico evolutionary biology professor Randy Thornhill responded, "...to say just generally that these new environments were more difficult, is open to criticism I think, because the tropical environment is terribly difficult, especially the diseases. People describe how they work in the tropics and they're just covered in parasites. It's like parasite rain." We need only reflect that the diseases of West Africa killed so many nineteenth-century European soldiers that it earned the sobriquet "White Man's Grave."

Eppig also disagrees with Kanazawa's insistence that the high disease rate in Africa (much of it infectious disease) can be caused by low human intelligence. "We know that the distribution of infectious disease is largely not determined by humans, it's determined by the diseases, and the types of environments that they thrive in. No one's ever caught a case of malaria in Alaska because the mosquitoes that carry it can't survive in Alaska.... That is entirely independent of human intervention."

Kanazawa's evolutionary biologist peers found his analysis sloppy and illogical. In spite of his unsupported claims, Kanazawa's racialized "blame the victim" approach to health harms is a strategy that industry has found useful in sidestepping responsibility for poisoning communities; it recurs often in these pages.

Hereditarian Follies

Perhaps Kanazawa's claims should be considered in light of his other writings that lack rigor but deliver shock value: For example, his poorly researched paid blog post for *Psychology Today* entitled, "Why Are Black Women Less Physically Attractive Than Other Women?," which refers to "race differences in intelligence" as if they were fact while elaborating on his inflammatory subject.[36] *Psychology Today* fired him, and Kanazawa apologized to the London School of Economics (but not to the women of African descent he insulted), admitting, "Some of my arguments may have been flawed and not supported by the available evidence." Despite my repeated e-mailed and telephoned requests over the course of three years, Kanazawa would not agree to speak with me or even respond to my queries regarding his perspective.

Hereditarian scientists who inspire far more reverence than Kanazawa have also embraced African American intellectual inferiority. From accounts in *Wired* to the *New York Times,* much is made of the fact that William Shockley, an old-guard hereditarian who is quoted above, was a Nobel laureate.[37] However, as a member of the Bell Labs research group in Manhattan he was jointly awarded the 1956 Nobel Prize in Physics for co-discovery of the transistor effect: His screeds on racial inferiority have naught to do with his celebrated achievement. Despite his ignorance of biology and genetics, he became a fervid eugenicist who advocated forcing or paying women of color to undergo sterilization in order to protect the gene pool. He wrote, "My research leads me inescapably to the opinion that the major cause of the American Negro's intellectual and social deficits is hereditary and racially genetic in origin and, thus, not remediable to a major degree by practical improvements in the environment."[38]

Perhaps Shockley's foray into racial calumny can be dismissed as an example of the "Nobel curse" in which laureates stray from their area of expertise to espouse risibly dubious beliefs. But another hereditarian Nobel laureate has far more credibility in genetics: James Watson, who shared the 1962 Nobel for discovering DNA's structure.

A genetic analysis by Iceland's deCODE revealed that James D. Watson, a Nobel laureate who claimed that African Americans are innately less intelligent than whites, has more "African" genes than the average European. (Public domain)

Watson's memoir entitled *The Double Helix* reveled in dismaying 1950s-era misogyny, and a sequel, *Girls, Genes, and Gamow,* updated his gender bias. But he took a break from his decades of hoary sexism in a 2007 interview with *The Telegraph* in which he declared his belief in African intellectual inferiority and theorized that black people have higher libidos.

He was, he said, "inherently gloomy about the prospect of Africa" because "all our social policies are based on the fact that their intelligence is the same as ours—whereas all the testing says not really." There is a natural desire that all human beings should be equal, but, he said, "people who have to deal with black employees find this not true."[39] He also predicted that genes responsible for creating differences in human intelligence could be found within a decade: That was in 2007.

I suspect that Watson's "gloom" only deepened in the furor afterward, when despite his belated disclaimer that "there is no scientific basis for such beliefs," he stepped down under pressure from his directorate at Cold Spring Harbor Laboratories.[40] Almost immediately afterward, an analysis by Iceland's deCODE Genetics showed that Watson has sixteen times more genes of African origin than does the average white European.

Kari Stefansson, deCODE's CEO, equated Watson's genome with the African DNA complement found in someone with an African great-grandparent, and told the *New York Times,* "This came up as a bit of a surprise, especially as a sequel to his utterly inappropriate comments about Africans."

The fact that a Nobelist subscribes to racial theories of intellectual inferiority is important because in our era of scientific complexity and rampant specialization, the role of authority is key. Many people who adhere to racial inferiority theories do not understand genetics well, and they are understandably swayed by the credentials of its high-profile espousers. Appointments at respected academic institutions, popular publications that boast impressively dense charts, graphs, and columns of data (as the racially fictitious 1840 census did), and Nobel Prizes persuade us, sometimes more powerfully than evidence can convince us.

The late J. Philippe Rushton, a professor at the University of Western Ontario (UWO), is referred to with the respect due his station even though he conducted research seeking to correlate race, intelligence, and sexuality by stationing himself in Toronto's upscale Eaton Centre shopping mall in 1988. There he surveyed white, black, and Asian shoppers in detail about their penis size and how far they could ejaculate.

Perhaps he used mall shoppers because the university had barred him from using students in his research after he pressured and deceived them into answering the same questions.

Although his university criticized his Eaton Centre study as "a serious breach of scholarly procedure,"[41] he was undeterred, and thereafter his eugenic screeds cropped up often in newspapers and magazines as well as in academia.

In 2005 Rushton, who had grown up in South Africa and became the head of the eugenic Pioneer Fund, told the *Ottawa Citizen* that he blamed the destruction of "Toronto the Good" on its black inhabitants[42] and that equalizing outcomes across groups was "impossible." He followed this with a 2009 speech at the Preserving Western Civilization conference in Baltimore, organized for the stated purpose of "addressing the need to defend America's Judeo-Christian heritage and European identity" from immigrants, Muslims, and African Americans.

Some scientists criticized Rushton's crude methods and outmoded scholarship, most notably his UWO colleague professor of psychiatry Zack Cernovsky. In 2010 he and I discussed Rushton when we shared a dais at the American Association of Psychiatry conference in New Orleans. Cernovsky noted, "Scientific scrutiny shows that Rushton's methodology (e.g., measuring head circumference by tape as a substitute for IQ tests) and his generalizations from inadequate samples discredit his work....

his books have probably misled at least some physicians who may resort to discriminatory practices in their clinical decisions about black patients."[43]

Yet before his death in 2012, Rushton was a respected member of his profession. In addition to holding a full UWO professorship, he was a Guggenheim fellow as well as a fellow of the American, British, and Canadian psychological associations.

Science journalists wield even wider authority. They are trusted to interpret complex medical and scientific subjects for laypersons, and none are trusted more than those staffing the august *New York Times,* "the paper of record." In 2014, Nicholas Wade, former *Times* science editor, wrote a book entitled *A Troublesome Inheritance: Genes, Race and Human History,* in which he asserted that scholars have proven human races to be a biological reality driven by evolution that forged racial differences in economic and social behavior. Wade suggests these genetic differences tell us why some live in tribal societies and others in advanced civilizations, why African Americans are (supposedly) more violent than whites, and why Asians may be good at business. However, the book traffics in shoddy logic and tortured data, which were disparaged by scientists and reviewers alike.

The *New York Times* review by David Dobbs condemned it as a "deeply flawed, deceptive and dangerous book." But the book's ultimate embarrassment was its disavowal by the very scientists whose work Wade cited to make his case. More than one hundred stellar evolutionary biologists signed a letter decrying Wade's "incomplete and inaccurate account of our research," concluding "there is no support from the field of population genetics for Wade's conjectures."[44]

What We Talk About When We Talk About Race

In 1990, the United States boasted nearly three hundred races or ethnic groups as well as six hundred Native American tribes.

Hispanics had seventy additional categories.[45] Medical journals routinely address race as a variable to be corrected for, as the focus of a medical question, and even as a consistent element of patients' profiles.

But rarely is "race" defined in these journals, and it often is used illogically. Studies often equate race with a genetic profile without attempting any genetic analysis, as a study of the congestive heart failure drug BiDil did in 2005. Some studies refer to "black Americans" in opposition to Hispanic Americans, even though the latter can be members of any race. "Latino" and "Hispanic" are often treated as synonymous, although the former refers to land of origin and the latter to language.

(For the record, although the terms "race" and "ethnicity" are sometimes used interchangeably, race is typically used to refer to affiliations that are assumed to be based on shared physical characteristics, whereas ethnicity refers to nonphysical affiliations based on culture, such as a shared religion or language, which in the case of Hispanics is Spanish.)

The sloppy use of racial categories in popular and scientific discourse is a holdover from the "one-drop rule," which held that any complement of African ancestry was enough to confer some degree of African American status on a person, or more accurately, to bar him from whiteness. Early in our nation's history, these regulations varied from state to state, as different degrees of interracial mating produced mixtures such as "octoroons" and "quadroons." In many cases, slave owners fathered children with the African American women they owned. Laws dictated what percentage of white forebears was necessary to allow a person to be defined as "white," to be so labeled in the census, to vote, to marry a white person, and so on.

In reality, there has been a great deal of unacknowledged mating between black and white Americans that hides our true genealogies and renders our racial labels meaningless.

Much attention is paid to Thomas Jefferson's genes going

into Sally Hemings's progeny, for example, but Hemings's genes also enriched the white gene pool. Driven by racial abuse and marginalization, many people who looked "white" enough simply chose to shake off the racial fetters by becoming white. Hemings's and Jefferson's daughter Harriet, for example, married a white man with whom she had children and then disappeared into whiteness and subsequently was lost to history, as Barbara Chase-Riboud revealed in her 1979 fictionalized but historically meticulous account *Sally Hemings* and its 1994 sequel *The President's Daughter*.[46] More recently, historian Catherine Kerrison recounts Harriet's fate in her 2018 account *Jefferson's Daughters: Three Sisters, White and Black in a Young America*.[47]

I vividly remember hearing Berkeley professor Troy Duster's brilliant response to a patronizing question from a television interviewer about his ancestry, "I come from a long line of slave owners." He was alluding to these shrouded genealogies. One study calculated that nearly one of every three white Americans possesses as much as 20 percent African genetic inheritance yet looks white.[48] More than one in twenty African Americans possesses no detectable African genetic ancestry.[49]

In spite of this, medical writings often equate race with a particular genetic profile, although a person's supposed race and her genetics map very poorly onto each other. "If we were to select any two 'black' people at random and compare their chromosomes," writes Sharon Begley in *Newsweek*, "they are no more likely to be genetically similar than either would be when compared to a randomly selected 'white' person."[50]

Race is a social construction, and although it has medical consequences, these too are socially driven. They are not definitive or determinative. In short, your medical condition tends to reflect the race to which you are perceived as belonging.

Medical reports frequently ignore the fact that America is comprised of both black and white Americans who sometimes differ in their health profiles, and sometimes do not. The medical literature also sometimes fails to recognize the differences in health risks and status that face Hispanics and Latinos of widely varying ethnic groups. In fact, medical journals often investigate questions of race without ever defining the terms "black," "African American," or "Hispanic," even as they discuss what purport to be consistent genetic characteristics within any of these diverse groups.

Nonetheless, average IQ differences among some U.S. racial groups—the widely touted "IQ gaps"—have been decreasing for the past decades. This narrowing may be partly a response to improved educational opportunities, or it could result from a decrease in the rates of some toxic exposures, such as lead's removal from gasoline and interior paint. Both explanations point to social inequality, not genetics, as a driver of IQ inequities.[51]

As I have mentioned, the Hispanic-white IQ gap is crudely flogged by hereditarians like Jason Richwine to support a xenophobic political agenda. For example, Richwine insisted that not only a lower average IQ but also a failure to assimilate was reason enough to halt Hispanic immigration to the United States. "Failure to assimilate" is a nonsensical accusation to level at people who have been consistently barred from higher education, employment, and homeownership, and who are frequent targets of raids and sweeps intent upon challenging their right to a place in America, even after decades of peaceful, law-abiding residency.

Asian Achievement: A Double-Edged Sword

One could be forgiven for failing to see how Asians are harmed by IQ mythology, which tends to place some Asians at the pinnacle of intellectual ability.

Not all Asians are accorded high IQ scores: Those from poor nations of the developing world, and their descendants in the

United States, share the low-IQ fate of their global South peers. According to one scale, the East Asian cluster (Chinese, Japanese, and Koreans) has the highest mean IQ at 105, followed by Europeans (100), Inuit-Eskimos (91), Southeast Asians (87), Native Americans (87), Pacific Islanders (85), South Asians and North Africans (84), sub-Saharan Africans (67), Australian Aborigines (62), and Kalahari Bushmen and Congo Pygmies (54).

Moreover, those Asians who occupy the top strata of IQ rankings find this status to be a double-edged sword. It doesn't as much laud Asians as it uses them to denigrate other, darker people of color. As UC Hastings law professor Frank Wu, author of *Yellow: Race in America Beyond Black and White,* writes, "Asian Americans are brought into the discussion only for the purpose of berating blacks and Hispanics."[52]

The high intelligence and intellectual achievement of Asians may owe something to the Pygmalion effect. In a classic 1960 experiment, California teachers were informed that as a result of IQ test scores, certain students of theirs were found to be "special," with prodigious potential and the expectation of intellectual greatness. Accordingly, the grades of the children labeled "special" improved dramatically, and, when tested a year later, half of their IQ scores had risen by 20 points. In fact, these children had been chosen at random, and the improvements in their scores served to demonstrate the outsize role that teachers' expectations can play in a student's academic success.[53]

The high IQ of some Asian populations is part of the "model minority myth," Wu points out. However, it is possible to be seen as "*too* intelligent," especially when high-IQ claims inspire unreachable goals that cause stress and even suicide among young Asian aspirants. The perception of Asians as naturally intelligent also inspires jealous resentment, causing other groups to target Asians and see that they are "taken down a peg."

The Asian-white IQ gap is increasing as Asians' IQs rise more swiftly than those of whites (just as those of African Americans are rising more rapidly than whites'), and there are signs of a backlash, including evidence that some colleges hold Asians to higher standards than other applicants. Schools deny this: In 2015 Princeton fought—and won—a discrimination suit over affirmative action policies that plaintiffs believe benefit African Americans and Hispanics while penalizing Asian students.[54]

As this book went to press, Harvard University was also defending itself against a lawsuit claiming that its affirmative-action policies favor African American and Hispanic students while harming the admission chances of Asian applicants with higher grade-point averages and standardized testing scores.[55] The case, widely seen as "a battle over the future of affirmative action," is supported by the conservative group Students for Fair Admissions, whose sixty-five-year-old leader, Ed Blum, helped to design the lawsuit.[56] (In 2016, an earlier lawsuit filed by Blum failed to block affirmative action at the University of Texas at Austin.)[57]

However, in the magazine *Diverse Issues in Higher Education,* Emil Guillermo describes this battle over affirmative action as "our civil war" and points out that the situation is far more nuanced than Blum's claims and the headlines suggest. "Fortunately," Guillermo writes, "recent polls suggest the vast majority of Asian Americans support affirmative action."[58] He explains that grades and scores are only part of the metrics that determine acceptance to Harvard and that Asian students as well as African American, Hispanic, and white applicants often gain acceptance to Harvard by dint of exemplary character, experiences, and nonacademic triumphs.

For example, student Sally Chen, a daughter of middle-class immigrants who describes herself as "one of those less-than-perfect Asian Americans," testified that her less-than-stellar grades and scores were redeemed by a high school record of exemplary leadership and civic involvement. Guillermo notes, "Chen testified that if Harvard couldn't consider race she wouldn't be at Harvard now. 'There's no way in which my flat numbers and résumé could've gotten across how much of a whole person that I am.'"

Asians' high IQs are frequently yoked to declarations that they lack creativity, are crafty, clannish, and constitute an "inscrutable" group that is unwilling to fully integrate into U.S. society—the same accusation that Richwine levels at Hispanic Americans to argue against their inclusion as citizens. In *Yellow: Race in America Beyond Black and White,* Wu writes,

> Every attractive trait matches up neatly to its repulsive complement, and the aspects are conducive to reversal.... To be intelligent is to be calculating and too clever; to be gifted in math and science is to be mechanical and not creative, lacking interpersonal skills and leadership potential. To be polite is to be inscrutable and submissive. To be hard-working is to be an unfair competitor for regular human beings, and not a well-rounded, likable individual. To be family oriented is to be clannish and too ethnic. To be law abiding is to be self-righteous and rigidly rule bound. To be successfully entrepreneurial is to be deviously aggressive and economically intimidating. To revere elders is to be an ancestor-worshipping pagan, and fidelity to tradition is reactionary ignorance.[59]

The "failure to assimilate" claim against Asians is especially unfair because racial bias often results in a refusal by other Americans to recognize their citizenship. Their identity as Americans is frequently challenged. It is well known that Japanese Americans were interned during World War II on a racial basis, but South Asians have also faced deportation and been stripped of their citizenship rights. Even today, fourth-generation Americans of Asian descent are often asked from what country they hail or complimented on their good English.

"Slothful" and "Dull"

Not all stigmatized minority-group members are people of color.

American "poor whites" have long been treated like an ethnic group that suffers from congenitally low intelligence. In 1928 the Carnegie Corporation bankrolled an investigation into the "poor white problem in South Africa" and administered IQ tests to more than 16,500 children, 20 percent of them poor whites. The study found that "Poor white IQs were lower than average."

In August 2016, outraged Britons forced top English educator David Hoare to resign after he publicly ranted against the Isle of Wight, once an upscale tourist destination. Hoare hadn't flung the n-word or decried the presence of Britons of color; instead he had targeted a quite different group, calling the region an "inbred, poor white, crime-filled ghetto." This view of indigent whites as a troublesome ethnic vector has long persisted in the United States as well, perfectly epitomized by Huckleberry Finn's lazy, illiterate, alcoholic, and criminal father, "Pap."

During the Civil War, fascinated Northern officers often described the ignorant, "coarse," landless Southerners they encountered in war zones. One wrote of North Carolinian combatants, "I pity them; most are not intelligent, not half of them can read." Another officer described their physical type: "A most forlorn and miserable set of people: white, long, lean, and lanky with long yellow hair."[60]

Before they achieved full white identity in the United States, Italians, Irish, and others from the "wrong" regions of Europe were also despised as inferior dullards prone to sloth, violence, and alcoholism. The eugenics movement and intelligence testing came to scientifically validate such views, and eugenicists buoyed the National Origins Act of 1924, which barred immigrants from southern and eastern European countries as "dysgenic." Low IQ scores provided a supposedly biological underpinning for this stereotype of poor whites as unintelligent, lazy, and prone to criminality, as it has for African Americans and Hispanics.

Today's "poor white" group identity is widely reflected in mass media, from *Cops* to Eminem to Jerry Springer—and in the pages of *The Bell Curve,* which denounced not only African Americans but also "underclass whites" as genetically predisposed to lower intelligence and higher criminality.

However, a more sympathetic *Atlantic* article, published in September 2016, entitled "The Original Underclass,"[61] symbolizes the recent attention given to the health status and behavior of lower-socioeconomic-class whites. Their profile of low achievement and drug addiction now casts them as an ethnic "underclass." Using IQ as a departure point, pundits showcase its supposed side effects—sloth, violence, a tendency to addiction, and even abnormal physical types, so that a picture often emerges of malnourished, slothful, unhealthy, dirty, gun-toting alcoholics with low intelligence.

The reification of this group as an unintelligent and degraded race follows the same illogic used to invest African Americans and Hispanics with pathology. Powerful pressures of economics, social stratification, trauma, and discrimination conspired to lower IQ scores in lower-socioeconomic-status whites. Research to improve their health status, including their IQs, is a laudable goal, but we must take care not to slip back into damning eugenic stereotypes. We need more research into factors that compromise the intelligence of lower-socioeconomic-status whites, including studies that quantify their toxic exposures.

Social Construction, Biological Consequence

The treatment of "poor whites" as a racial group illustrates race is not an innate biological reality but a sociopolitical construct that is useful for maintaining political biopower.

But race is a social reality with real-world biological consequences and nowhere is this more apparent than environmental racism. De facto racial segregation, mortgage redlining, and, as I shall describe, the withholding of basic environmental services, are used to force racial groups into environmental "sacrifice zones," where exposures to high levels of IQ-lowering heavy metals, chemicals, and pathogens impede normal brain development. Targeting the groups with aggressive marketing of especially noxious alcohol and tobacco products and confining them to "food deserts" that help to ensure brain-eroding nutritional deficiencies all have marked biological consequences. When it comes to exposures that limit cognition, race as a social construction becomes race as biological fate. Unless we choose to intervene.

We could not do better than to start with lead.

PART II

The Brain Thieves

CHAPTER 2

The Lead Age: Heavy Metals, Low IQs

If you were going to put something in a population to keep them down for generations to come, it would be lead.

— Detroit pediatrician Mona Hanna-Attisha[1]

Anthony Miller, ten, could not sit still.* Ensconced in the steely mesh of an Aeron chair at the head of a conference table ringed with lawyers, his eyes darted about the room.

Anthony glanced quickly at the camera recording him, then looked down nervously, then up again, at his mother and her lawyer. Shifting in his seat, he worried that he would somehow give the wrong answers at what seemed like an important meeting, although everyone had assured him he need only do one thing: tell the truth.

After a few moments, the lawyer asked, "If your mother were to win the case and you had plenty of money, what would you want most?" Abruptly, Anthony sat up straight, looking intently at the lawyer.

* Anthony is not his real name. Other deposition details have been changed to protect his privacy.

"I want to read," he said with audible fervor. "I just want to be able to read."

Anthony is one of at least 37,500 Baltimore children who suffered lead poisoning between 2003 and 2015. Nearly all were African American.

Lead poisoning is often caused by exposure to industrial emissions, tainted water, or poorly maintained pre-1978 housing that features flaking lead paint and lead dust. Nearly two of every five African American homes are plagued by lead-based paint.

Trojan Horse: Lead Pervades Baltimore

One of these children was Ericka Grimes.

The young woman fitted the key of their new apartment into the gleaming brass lock and turned it smoothly, with a satisfying click. She smiled up at her husband: already better than their old house, where the key often got stuck or jammed in the lock, slowing her down when she had an armful of groceries. It's true that the house itself was weathered and a bit shabby, but the young mother cared most about what it *didn't* have: lead, a toxic metal common in Baltimore's African American neighborhoods.

She knew that children sickened and died as a result of growing up in lead-tainted housing, and she was determined to avoid that fate for her children. She was therefore encouraged when, three years after moving in, she was approached by the Kennedy Krieger Institute (KKI) to participate in a lead-abatement study at her property. On its website KKI described itself as a place where she could access an "interdisciplinary team of experts in the problems and injuries that affect your child's brain, and receive personal compassionate care for your child." Moreover, it is affiliated with the prestigious Johns Hopkins University.

When the KKI drew her daughter Ericka's blood on April 9, 1993, her reading was a reassuring nine micrograms of lead per deciliter of blood (μg/dL), which at the time was a "normal" reading according to CDC guidelines. (These were later revised. Experts now consider no level of blood lead safe.)

When Ericka was retested the following September 15, her blood-lead reading had shot up to 32 μg/dL, a "highly elevated" reading that is six times higher than the allowable limit set by the CDC.

Two-year-old Ericka had fallen victim to the mid-1990s "Repair and Maintenance Study" in which KKI researchers undertook a study of African American families who lived in 108 units of decrepit housing with interiors that were encrusted with crumbling, peeling lead paint.

Lead paint is a notorious cause of acute illness and chronic mental retardation in young children. Among its many signs and symptoms are slowed growth, anemia, heart disorders, reproductive problems, reduced kidney function, lowered IQ, and learning and behavioral difficulties. Unfortunately, these take time to show themselves, and it is virtually impossible to tell that an infant like Ericka has suffered lead poisoning by looking at her: a medical evaluation, including a blood test, is needed.

Like many toddlers, Ericka had been poisoned when she inhaled airborne lead dust and ingested lead dust by putting her hands in her mouth after crawling on the floor. Perhaps she nibbled the peeling paint chips from windows and doorjambs, drawn by the sweet taste of the lead.

That same sweet taste led Romans to infuse their gastronomic delicacies and wine with lead, courting the mental devastation that some historians believe hastened the civilization's decline. The lead used in everything from their cosmetics to the pipes in the empire's famous aqueduct system did not help matters. Even the word for such systems, "plumbing," derives from the Latin word for lead, *plumbum,* as does the chemical symbol for the metal, Pb.

For us, as for our forebears, lead's sweet utility has proved disastrous, and we are suffering the same stultifying fate, although today it is not jaded Roman epicures but poor African American children in crumbling inner-city housing who suffer most from lead.

Why? It's not as if we don't know lead's horrors. We now understand that we can protect children only by banning the use of lead paint and by offering lead-abatement programs that completely remove the remnants of the toxic paint and dust. As early as 1987 the CDC advocated complete abatement because there is no safe level of exposure.[2]

Yet the city of Baltimore abounds with lead-tainted low-income housing populated almost exclusively by African Americans. In Baltimore, statutes requiring that lead paint be thoroughly abated from home interiors have made renting lead-tainted housing illegal.

But Baltimore slumlords find removing this lead too expensive and some simply abandon the toxic houses. Cost concerns drove the agenda of the KKI researchers, who did not help parents completely remove children from sources of lead exposure. Instead, they allowed unwitting children to be exposed to lead in tainted homes, thus using the bodies of the children to evaluate cheaper, partial lead-abatement techniques of unknown efficacy in the old houses with peeling paint. Although they knew that only full abatement would protect these children, scientists decided to explore cheaper ways of reducing the lead threat.

So the KKI encouraged landlords of about 125 lead-tainted housing units to rent to families with young children. It offered to facilitate the landlords' financing for partial lead abatement—only *if* the landlords rented to families with young children. Available records show that the exposed children were all black.

KKI researchers monitored changes in the children's health and blood-lead levels, noting the brain and developmental damage that resulted from different kinds of lead-abatement programs.

These changes in the children' bodies told the researchers how efficiently the different, economically stratified abatement levels worked. The results were compared to houses that either had been completely lead-abated or that were new and presumed not to harbor lead.

Scientists offered parents of children in these lead-laden homes incentives such as fifteen-dollar payments to cooperate with the study, but did not warn parents that the research potentially placed their children at risk of lead exposure.

Instead, literature given to the parents promised that researchers would inform them of any hazards.[3] But they did not. And parents were not warned that their children were in danger, even after testing showed rising lead content in their blood.

This shocking research violated a number of ethical rules. To protect the safety and rights of research subjects, every federally funded human experiment must be approved by an institutional review board, or IRB, which is the institution's principal body charged with protecting the subjects of medical research.[4] One might wonder how the study cleared this hurdle, given the plan to maintain children's exposure to a known toxic metal for research that was driven by economic issues and was nontherapeutic, providing no possible benefit to them. The IRB of the prestigious Johns Hopkins University, with which the KKI is affiliated, approved the protocols after helping the researchers refine their application in line with governmental requirements. Although the Office for Human Research Protections (OHRP) within the Department of Health and Human Services ostensibly provides governmental oversight of IRBs,[5] routine scrutiny is nearly nonexistent. Many modern human studies have been approved and conducted although they are ethically flawed, sometimes deeply so.[6]

Ericka was exposed from her birth in 1992 until 1994, when her parents discovered that she was profoundly lead poisoned, and the family moved out. Their subsequent lawsuit against KKI alleged that the researchers were aware of the lead-paint hazard and that the KKI violated its duty to the children by not fully explaining the dangers of lead paint in the consent form that the families signed in order to participate in the study.

The parents of children in a similar study were given a consent form in which the KKI did not clearly disclose that the children might accumulate dangerous levels of lead in their blood as a result of the experiment, nor did it describe the effects of elevated blood lead, including damage to the central nervous system, irreversible behavioral problems, and death. Grimes's parents also claim that the KKI "failed to warn in a timely manner or otherwise act to prevent the children's exposure to the known presence of lead."[7]

The KKI countered that its researchers had no contract or "special relationship" with the study subjects and owed no duty to them. The circuit court agreed. Ericka Grimes had lost.

But on August 16, 2001, Maryland's top appellate court reversed this decision and ruled against the researchers. In doing so, it drew a parallel between the Grimes case and the Tuskegee syphilis experiment, a forty-year study in which hundreds of African American men in Alabama were lied to and maintained in an infected, ill condition by U.S. Public Health Service researchers who failed to treat their syphilis. The USPHS did this in order to study the disease's ravages on the men's bodies at autopsy, just as the bodies of the lead-exposed children were used to gather data.[8]

The appellate court found using the children as biologic monitors ethically indefensible, as the language in Judge R. Cathell's opinion made clear. He wrote, "It can be argued that the researchers intended that the children be the canaries in the mines."

Worse, as his decision noted, the children had been abandoned by the IRB that was charged with protecting them as research subjects: "The IRB was willing to aid researchers in getting around federal regulations designed to protect children used as subjects in nontherapeutic research. An IRB's primary role is to assure the safety of human research subjects—not help researchers avoid safety or health-related requirements."[9]

Ericka's family eventually reached a settlement of its claims against KKI, but twenty-five other parents in the study filed a class-action suit against the Kennedy Krieger Institute in September 2011, accusing it of negligence, fraud, and battery as well as violations of Maryland's consumer protection act. At the time of writing, no decision has been made on the case.

Although I am not involved in the class action, over the past decade I have worked as a consultant to law firms that represent children who, like Ericka, were sickened by lead in the Baltimore studies. Many researchers have defended such studies, and one of the stranger defenses I have heard, albeit only from one lawyer, is that because virtually all the children living in Baltimore's lead-imbued housing were African American, maintaining them in toxic housing during the study was not something for which researchers should be blamed. This defense is not only patently ridiculous but also ignores the fact that some of the children poisoned in the experiment had *not* previously suffered dangerously elevated lead levels.

Such a claim of blamelessness is belied by the fact that the researchers were not objective, innocent bystanders. They actively encouraged landlords to rent to families with vulnerable young children by offering financial incentives.

The participation of a medical researcher, who is ethically and legally responsible for protecting human subjects, changes the scenario from a tragedy to an abusive situation. Moreover, this exposure was undertaken to enrich landlords and benefit researchers at the detriment of children, so it also evokes the shade of British philosopher Jeremy Bentham who wrote, "No man ought to take advantage of his own wrong."[10]

Heavy-Metal Mayhem

How did lead become such an important hazard, and how did it come to preferentially threaten the minds of African Americans and other communities of color?

As early as the 1880s, scientists realized that lead plumbing was poisoning our water, like that of the Romans before us. By the 1920s many cities and towns passed statutes that banned or sharply restricted the use of lead pipes, but the lead industry, notably the Lead Industries Association (LIA), pushed back. The LIA's vigorous "educational" campaign sought to rehabilitate lead's image, muddying the waters by extolling the supposed virtues of lead over other building materials. It published flooding guides and dispatched expert lecturers to tutor architects, water authorities, plumbers, and federal officials in the science of how to repair and "safely" install lead pipes. All the while LIA staff published books and papers and gave lectures to architects and water authorities that downplayed lead's dangers.[11]

Over the succeeding decades, the hazard was gradually forgotten, an early exercise in industry's skill in manufacturing doubt regarding environmental poisoning. It would not be the last such success, and as a result, the contribution of lead pipes to the nation's lead poisoning crisis continues in Flint, Michigan, and beyond. Meanwhile, lead-tainted water gradually took a backseat to what would prove its most efficient route of childhood poisoning—leaded gasoline.[12]

In the 1920s, when General Motors (GM) considered adding its patented chemical tetraethyl lead (TEL) to gasoline, it already had a cheap, ubiquitous, perfectly serviceable anti-knock[13] compound for high-compression engines: ethanol, the same alcohol found in the beer, wine, and hard liquor that we drink.

Some beneficial effects of using ethanol fuel were recently illustrated when the airborne-lead concentration of São Paulo, Brazil, plummeted 65 percent between 1980 and 1985—and so did its notoriously high crime rate. In a bid to reduce its dependence on imported oil, Kevin Drum explained in *Mother Jones,* Brazil began substituting lead-free E95 fuel ethanol for vehicles in 1979. By 1987, the ethanol fueled nearly half the vehicles in Brazil and a much higher percentage in São Paulo, where the annual homicide rate fell from over 12,000 in 2000 to less than 5,000 in 2012. (Crime rates, Drum notes, typically plummet about twenty years after the drop in lead exposure, when the data of a new, unpoisoned generation are tallied.)[14]

But despite its safety, alcohol's very cheapness and ubiquity worked against it in GM's eyes. Alcohol was in common use and people even brewed it in home stills, so it was not "novel" and therefore could not be patented. This meant GM could not profit from its exclusive sale, nor could it hope to corner the automotive market. Moreover, Standard Oil declared itself "reluctant... to encourage the manufacture and sale of a competitive fuel [that is, ethanol] produced by an industry in no way related to petroleum."[15] Standard Oil's dislike of the ethanol additive was a powerful disincentive for GM because it sought to partner with it and other energy behemoths like DuPont.[16]

So, in yet another example of industrial greed trumping public safety concerns, GM chose to use lead as an anti-knock additive in a 4:1 mix of gas to TEL, despite the fact that lead was costlier, less readily available, and, as GM knew from the beginning, "very poisonous."

It was March 1922 when, as Kevin Drum writes, "Pierre du Pont wrote to his brother Irénée du Pont, Du Pont company chairman, that TEL is 'a colorless liquid of sweetish odor, very poisonous if absorbed through the skin, resulting in lead poisoning almost immediately.' "[17]

Subsequently, gas emissions from cars and trucks drove up lead levels in the air alarmingly and permanently. Because inhalation is the most efficient route of poisoning, blood-lead levels rose around the country, especially quickly in children, who have proportionately more lung-surface area than adults.

Despite the 1922 caveat by DuPont, companies later denied—repeatedly—knowing that leaded gasoline was poisonous.

Lead may be the cause of the biggest childhood poisoning epidemic ever, because even today when we know the dangers of only minimal exposure, lead abounds in more than paint and gas. Lead lurks in battery cases, building materials, burial vault liners, lead crystal, pewter, solder, shielding, computer monitors, ceramic pots, television components, and even modern makeup. It is the most widely dispersed environmental toxin affecting children in this country.

As with many environmental poisons, the affected communities are often characterized as "poor," or of low socioeconomic status. But this language is imprecise, and it veils the heavily racial nature of exposures; this is an epidemic afflicting African American and Hispanic children as well as the poor.

About 39 percent of African Americans and 43 percent of Hispanics are poor, and some definitions of "socioeconomic" already subsume race as a factor. This means that race becomes an endemic but hidden characteristic of the affected neighborhoods. Also, middle-class blacks are far more likely than middle-class whites to suffer from environmental hazards placed in their communities. The 2014 report *Who's in Danger: A Demographic Analysis of Chemical Disaster Vulnerability Zones* notes that middle-class African American households with incomes between $50,000 and $60,000 live in neighborhoods that are more polluted than those where very poor white households with incomes below $10,000 live.[18]

The failure to identify the imperiled groups as racial minorities may reflect a level of discomfort in acknowledging America's racial disparities. These include the persistence of racial segregation and the slow violence of racial environmental poisoning. It may be instructive to remember that despite the fact that Flint is 57 percent black, the lead-poisoning crisis in the city was first described in 2014 as affecting a "poorer community" or a lower socioeconomic group. Only in 2016, the year the Center for Effective Government revealed that children of color constituted almost two-thirds of the 5.7 million children who live within a mile of a toxic facility, was the Flint lead crisis routinely described as a racial assault affecting African Americans.[19]

Descriptions of the vulnerable that focus on economics and exclude race are semantic shrouds that are not only euphemistic but inaccurate. Poverty is certainly a risk factor for environmental exposure, but race is a larger and more consistent one.

Industrial Secrets

This wasn't always the case with lead poisoning. Lead was widely introduced into American homes for the same reasons the ancient Romans embraced it wholesale: it was useful, and it enhanced the home environment with higher-status ornaments and products.

Painted walls, for example, replaced wallpaper in midcentury homes because paint was more modern, easier to clean, and therefore more hygienic, as the industry reminded us. As TEL use in gas declined, the lead industry offset its losses by marketing lead pigments. It used powdery lead carbonate, which was mixed into oil-based paint to brighten colors and make the paint more durable and washable. Lead-based paint was used widely in schools, hospitals, and other public buildings.[20]

The lead industry knew that the paint, like TEL gas, was very poisonous.

As early as 1900, a Sherwin-Williams newsletter declared that lead was a "deadly cumulative poison."[21]

In 1904, Sherwin-Williams's own in-house magazine described lead as "poisonous in a large degree, both for workmen and for the inhabitants of a house."

By 1912, the company National Lead reported to shareholders that women and children were forbidden from working on its lead products because of the devastating effects—all while the company continued to manufacture lead-based goods for home use, including children's toys.

By 1955, the LIA health and safety director Manfred Bowditch wrote an internal memo, lamenting, "With us, childhood lead poisoning is common enough to constitute perhaps my major 'headache.'"[22]

So the poisonous nature of lead paint was well known from the beginning. Safe, lead-free alternatives like ethanol and unleaded house paint were also available from the beginning. But the LIA's aggressively deceptive marketing efforts in the early twentieth century made lead paint popular among homeowners and landlords.

Lead dust from the paint coated everything, including the lungs of children. When children become toddlers, able to move about and explore their poisoned world more freely, the real devastation began. Young children's principal means of exploring the unfamiliar world is to put objects in their mouths. Even noxious tastes won't discourage young children from mouthing objects, but lead's sweetness encouraged them to lick and suck on painted cribs and windowsills, and to frequently put their unwashed hands in their mouths after peeling sweet paint chips off household surfaces.

This is a very efficient method of poisoning. A lead-paint chip no larger than a fingernail can send a toddler into a coma and death. One-tenth of that amount will lower his IQ.

The lead industry's response? Rather than pull lead-based

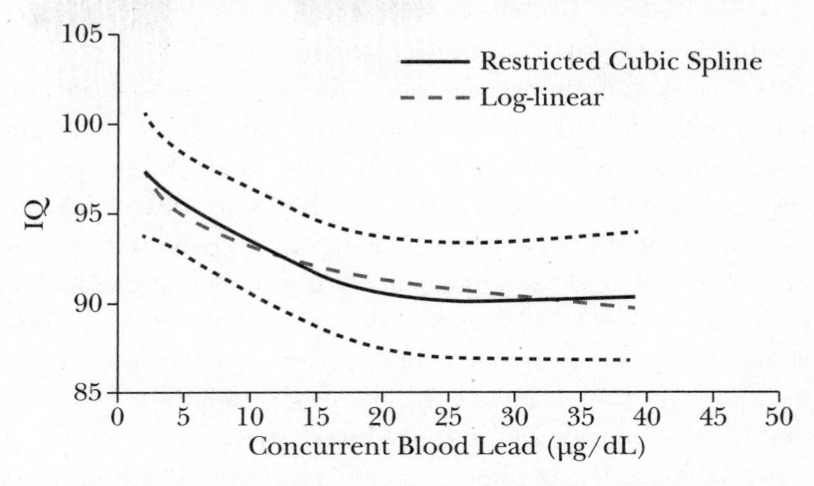

Lower IQs are found among children with the highest blood-lead concentrations.

paint from the market in favor of safe alternatives, firms more aggressively touted lead paint as a healthy wall covering. Because it was so easy to wash, it was specifically recommended for children's rooms. Lead pigment maker National Lead Industries, the same company that wouldn't allow women and children workers in their lead paint factories due to safety concerns, ran a *National Geographic* ad in 1923 that claimed, "Lead helps to guard your health." The following year, an advertisement by Sherwin-Williams boasted that after her home received a coat of lead paint, "Cousin Susie says her health improved instantly."[23]

In the 1990s, tobacco companies came under fire when Philip Morris sought to woo customers from an early age with trinkets like child-sized flip-flops that stamped the word "CAMEL" in the sand and cartoon characters like Joe Camel, which the *Journal of the American Medical Association* found was as recognizable to six-year-olds as Mickey Mouse. They were taking a page from the book of the lead industry, which marketed lead products designed specifically for children: lead-painted toy horses, train sets, and jewelry—even coloring books that celebrated the superior colors of leaded paint as vibrant hues that are perfect for the cheerful decor of children's bedrooms.[24]

Of course, both chronic and acute exposures to lead poisoned industry workers, too. One of the more dramatic examples of this occurred on October 27, 1924, when eleven workers were poisoned by TEL gas at the Standard Oil research laboratory in Elizabeth, New Jersey. They suffered profound confusion, delirium, and "derangement" so dramatic that one had to be bound in a straitjacket before being hospitalized, where one worker died. Such mental symptoms were well known to the workers, who called the site the "loony gas building."[25] Similar outbreaks of poisoning, and madness among three hundred other workers who became psychotic and some of whom were also carried away in straitjackets, took place at the other two TEL manufacturing sites. Between three and fifteen workers died there, according to University of Pittsburgh professor Herbert L. Needleman, a preeminent lead-poisoning researcher.[26] The Elizabeth, New Jersey, disaster spurred the beginning of restrictions on leaded gasoline, whose levels were progressively reduced by laws passed over the decades until 1974, when the Environmental Protection Agency required oil companies to stop putting lead in gasoline. In 1978, lead paint was banned in new home construction, although nothing barred its use in older and existing housing.

Eliminating lead from gas resulted in quantifiable and dramatic health gains. The amount of lead in Americans' blood fell by four-fifths between 1975 and 1991.[27] Fewer children fell into coma and death from lead inhalation, and as Kevin Drum of *Mother Jones* points out, the nation's IQs even rose in the aftermath of lead's banishment from gas.[28]

Public-health advocates campaigned to eradicate lead from

paint, too, but the lead industry fought to retain its lucrative market with tactics similar to those that had allowed it to retain lead plumbing in the face of frank poisoning. The LIA successfully lobbied key lawmakers and public health officials, including scientists, to look the other way. Its viability was at stake, especially because some of its members, like the National Lead Company and Eagle-Picher (formerly Eagle White Lead), manufactured both lead paint and lead pipes.

Richard Rabin, M.D., recalls,

> In 1933, the Massachusetts Department of Labor planned to prohibit the use of lead paint inside homes, due to the danger to young children. However, the Lead Industries Association, concerned that such a regulation could lead to the banning of lead paint in other states, persuaded the Labor Department to drop the rule. Instead, the Department issued a weak recommendation that lead paint not be used where children could have access to it, without reference to lead's toxic nature.[29]

In 1956, LIA health and safety director Manfred Bowditch boasted to the Secretary of the Interior, who was responsible for regulating the lead industry, that "with the public health officials local, state, and national I been [*sic*] at some pains to cultivate their good will and get them into a receptive frame of mind as to our viewpoint. I feel this has paid off, as for example, in Chicago where we have been able to stave off a paint-labeling regulation like that here in New York."[30]

As David Rosner and Gerald Markowitz reveal in their 2013 book *Lead Wars: The Politics of Science and the Fate of America's Children,* the industry consistently denied that lead paint was a danger to children. Internal communications from the LIA imply that they seemed most worried about the economic effects of lead's toxicity and about their public image.

> In the first place it means thousands of items of unfavorable publicity every year. This is particularly true since most cases of lead poisoning today are in children, and anything sad that happens to a child is meat for newspaper editors and is gobbled up by the public.
>
> It makes no difference that *it is essentially a problem of slums, a public welfare problem.* Just the same the publicity hits us where it hurts.[31]

Landlords who didn't relish the prospect of expensive lead abatement and repainting also clung to such denials of responsibility.

The Uses of Doubt

The LIA and its scientists also churned doubt regarding lead's dangers by minutely challenging data that demonstrated lead toxicity.

This pattern has been repeated consistently in the evaluation of environmental poisoning. Raising—and where necessary, creating—doubt in response to health data is a favored ploy of the industry and of the scientists it employs and funds. As public health researchers and practitioners amassed data showing sickness and death, the industry contested their studies and demanded ever-increasing levels of surety before any abatements or eradication commenced.

For the lead industry, as for the industries behind many subsequent poisoning cases, doubt became a useful foil against the expense of regulation. As the title of one book on the subject proclaims, *Doubt Is Their Product.*[32]

This corporate skepticism is often articulated as a scientific question, to wit, "Is there really incontrovertible evidence that lead in paint is a hazard demanding eradication?"

Demanding scientific proof sounds logical before undertaking an expensive ban and an even more expensive removal and rehabilitation of a toxic site. Some think that the expense and disruption demand certain proof. But finding such proof is very expensive, too: research requires a great deal of funding and other resources, and it can take decades to amass data that can utterly condemn or exonerate a suspect chemical.

But the most precious resource lost in the long decades of research required for absolute proof of a chemical's harmfulness is health: the lives of the people who are sickened, killed, or hobbled by lowered intelligence during the long, expensive search for the ever-higher standards of proof demanded by industry and its scientists. "Its scientists" refers not only to those in its employ, but to those whose work it funds, because a wealth of studies show that industry-funded research, even that published in top-tier journals, tends to find results that buttress the interests of the industry that pays for them.[33] Doubt was and is more than a scientific question: it is often a profitable stance as well.

Am I suggesting that chemicals suspected of causing illness be restricted or removed from the market even before we can prove that they cause the sickness, including memory loss, IQ loss, confusion, and behavioral problems suffered by those who are exposed to them? Yes.

As early as 1988, the Agency for Toxic Substances and Disease Registry (ATSDR) estimated that the blood lead levels of 2.4 million U.S. children and 400,000 fetuses exceeded the blood-lead-level standards of the day, which were then quite high, at 15 μg/dL. This is a fraction of the millions who have fallen into comas, died, been permanently injured, or lost intellectual capacity as a result of lead poisoning while the industry successfully blocked regulation and bans by simultaneously hiding damning data and demanding incontrovertible proof that lead was injuring Americans. These life-changing illnesses and deaths cost victims and their families. But they also cost the nation a great deal of money for the victims' medical care and their dependents' care. And this widespread loss of intelligence costs the nation the brainpower it needs to compete in the modern industrialized economy. Even an average IQ loss of five points has been shown to shrink the pool of gifted U.S. children and to swell the ranks of the mentally incapacitated.

Unless we want to repeat this scenario with pesticides, PCBs, mercury, phthalates, and other contemporary toxic exposures, we must react much more quickly with legislative bans and protection for people on the front lines of environmental poisoning.

But I am not suggesting that we neglect proving the danger, or safety, of industrial chemicals. This is imperative. The problem is that we are testing them at the wrong time.

The United States has approved 60,000 chemicals for industrial uses that expose Americans in the workplace, in fence-line communities, and in company towns, and that poison us by leaching into our water and air. Regulations do not require companies to test the effects of such chemicals on humans prior to using them, and they don't. We typically learn of health hazards only after people are exposed to them, and because environmental pollution exposes victims to a mixture of untested substances, teasing out and accurately characterizing the effects of any one chemical can be difficult. Moreover, many pollution-caused illnesses, such as lung cancer or mesothelioma, can take years or decades to appear, making them difficult to connect to a given chemical exposure.

But some nations, like those of the European Union, test industrial chemicals differently, *before* they are used. So should we. Rather than collecting, analyzing, and haggling over the sufficiency of data about a chemical's relative safety *after* allegations of harm are made, the safety of any potentially toxic substance should be demonstrated *before* it is marketed and used.

Determining the safety of industrial chemicals before they

go into use is an illustration of the *precautionary principle.* Unfortunately, it is not favored in the United States. And although it is not favored by corporations because it would add to the expense of research and development and delay marketing, pre-market testing in line with the precautionary principle would also save them (and the nation) the expense of bans, cleanups, and lawsuits. Most importantly, it would save the lives, health, and intellect of millions of poisoned Americans each year. As Swedish filmmaker Bo Widerberg once wrote, "Sometimes it doesn't make sense to ask what things cost."

It could, however, make a great deal of economic sense to make sure these chemicals are safe by testing them before putting them on the market, given the astronomical cost of health care and lost wages due to poison-induced illness—not to mention the expense of cleanups.

The lead industry relentlessly—and successfully—questioned the science showing that its products were killing children in their own homes. Industry so often resorts to the tactic of introducing doubt that, as later chapters in this book document, it has successfully halted or delayed the restriction and banning of clearly toxic substances.

For more than a half century the LIA internally acknowledged lead's toxicity while publicly casting doubt on its dangers and blaming victims by ascribing illness to "improper handling."

Meanwhile, public health practice focused on institutional accountability to ensure or improve American health. Its researchers, institutions, and government agencies sought to cajole, convince, or force organizations to comply with standards. Pressure was brought to bear, and eventually, the industry had to conform to lead-removal standards in order to minimize children's exposure.

But as Columbia University historian and public health professor David Rosner and his coauthors point out in their article "The Exodus of Public Health," a sea change in philosophy was taking place. Public health pressure on industrial polluters has been largely supplanted by a focus on laboratory science and the mantra of "personal responsibility."[34] "The field...abandon[ed] universalist environmental solutions—introducing pure water, sewage systems, street cleaning—and begin focusing on training people how to live cleaner, more healthful lives," they write. By the advent of the Cold War, science and medicine became great levelers, allowing public health professionals to ignore social factors—including the racial segregation, poverty, inequality, and poor housing that had been the traditional foci of public health reformers only thirty years before—and explain disease without any of the disruptive implications of a class analysis.

By the 1970s, a powerful discourse of personal responsibility for health and disease placed blame on individuals and implicitly absolved corporations that marketed harmful products such as cigarettes and lead paint and polluted the nation's water and air.[35]

The pressure to improve health is now placed on individuals as their behaviors are criticized, scrutinized, and sometimes demonized, in the hopes of modifying their behaviors into healthier ones.

Personal responsibility for health is important and a good concept in theory. However, a difficulty arises when avoiding disease requires avoiding lead fumes and dust, shunning tainted tap water, or evading mercury emissions from coal-fired power plants, because individuals have little power to control their exposure to industrial pollutants that frequently poison their workplace and neighborhoods.

Blame the Victim

When muddying the scientific waters no longer worked to hide how lead was killing and intellectually crippling children, the lead industry resorted to another tactic—deflecting blame onto the injured.

A wider realization of lead's hazards coincided with a demographic upheaval in the aftermath of 1960s civil rights–era violence. This friction included rebellions, race riots, and the legally mandated integration of schools and housing. As whites fled to the suburbs, people who could not do so, primarily African American and Hispanic families, remained trapped in the cities' lead-tainted housing. Lead exposure was morphing into a racially disparate poison.

The industry found this helpful to their cause as the purveyors of lead shifted the blame from their own promotion of a toxic product to a more nebulous malefactor: poverty.

In an internal memo Manfred Bowditch, the health and safety director of the LIA, wrote, "Aside from the kids that are poisoned (and we still don't know how many there are), it's a serious problem from the viewpoint of adverse publicity. The basic solution is to get rid of our slums, but even Uncle Sam can't seem to swing that one."[36]

In blaming manufactured racial disparities on "poverty," Bowditch used what is still a popular ploy to evade industrial responsibility. In *Deadly Monopolies: The Shocking Corporate Takeover of Life Itself,* I discuss a public relations strategy in which pharmaceutical manufacturers fought bad press in the wake of the Agreement on Trade-Related Aspects of Intellectual Property Rights (TRIPS) regulations negotiated by the World Trade Organization (WTO) in 1994.[37] These laws, pushed through the WTO by the industry, force developing nations to adhere to Western pharmaceutical patents—and to accept high Western prices. These patents protect high corporate profits, but allow pricing above what the populace of countries like Thailand, Sudan, the Democratic Republic of the Congo, and Brazil can afford.

They also allowed companies to abandon the development of medications they deem unprofitable, often medicines intended for diseases of the developing world. The TRIPS regulations even prevent countries like India from making cheap similar drugs that once helped to fill the pharmaceutical void in the developing world.

But pharmaceutical-company spokespersons and their scientists insisted that "poverty, not patents" was separating people in the tropics from the medications they need, often in articles that used quite similar wording and were based on deeply flawed studies.[38]

Similarly, when the LIA suggested that the nation could only eliminate lead poisoning if it managed to "get rid of our slums," they were washing their hands of responsibility, as if leaded gas, lead paint, and lead-coated toys had spontaneously generated within slums rather than having been aggressively placed there via the industry's vigorous promotion and political lobbying.

As the LIA statement shifted the blame onto poverty, for which it couldn't be blamed, it portrayed lead poisoning as just another pathology of black Americans, and therefore of little concern to most Americans.

In addition, the LIA narrative argued that regulating lead was futile. As the industry knew, no one saw getting rid of slums as a feasible solution: "even Uncle Sam can't seem to swing that one."

But more than poverty trapped black and Hispanic families in lead-poisoned homes. Limited educational and career opportunities drove that poverty, and "slums" were the product of racial segregation. The law had once encoded segregation, but even after "separate but equal" housing legislation was revoked, racial and economic bias maintained housing segregation. Like most Americans, the industry understood that the nation was not eager to allow African Americans out of the slums and into white neighborhoods en masse, no matter how much handwringing attended the discussion of ghetto problems. In short, the industry's stance that lead poisoning was "a public welfare problem," not a lead-regulation problem, placed it far from the industry's purview.[39]

But the LIA's disclaimers revealed even more: a fresh blame-the-victim tactic. The term "slum" suggested the racial component of the problem, but Bowditch continued with language that made the racial characterization explicit. Lead was safe when applied and used properly, the industry suggested, but it implied that parents of color were slovenly housekeepers who did not use the paint in a correct, "safe" manner, who allowed unsafe levels of lead to accumulate in their dusty, dirty homes and did not monitor their children's proper use of lead-painted toys and products. Of course this message still entailed a tacit acknowledgment that lead is dangerous, but the powerful invocation of ethnic parents as the real problem seemed to overshadow this misstep.

"Next in importance is to educate the parents," Bowditch wrote, "but most of the cases are in Negro and Puerto Rican families, and how does one tackle that job?"[40]

"Uneducable" Negroes and Puerto Ricans

In casting black and Hispanic parents as uneducable, Bowditch tapped into a perception of people of color as too unintelligent to keep themselves and their children healthy and safe. He indicts black and Hispanic intelligence, not his firm's aggressive marketing and duplicitous defense of toxic wares.

This mixture of corporate half-truths and racist blame-the-victim characterizations helped create an image of lead poisoning as a racial ailment that was of little concern to the average American.

This venal tactic resonated with many Americans, in part because by the 1970s lead poisoning was indeed a racial problem.

One might think that the end of de jure segregation would

have made lead exposure democratic by yielding racially integrated neighborhoods and risks. Unfortunately, de facto segregation persisted, leaving people of color to deal with lead-imbued housing in the aftermath of white flight to the newly constructed, lead-free developments of the suburbs. Even middle-class people of color were affected. Redlining, credit inequities, educational and employment discrimination, and lower wages ensured that most marginalized minority-group members could not penetrate the pristine, pollution-free suburbs. But neither could African American doctors, lawyers, and corporate executives of color. They were barred by race, not income, from suburban homes not only in Southern states, but throughout the country. In upstate Rochester, New York, for example, Eastman Kodak chemist and executive Dr. Walter E. Cooper documented for the local newspaper his long odyssey to buy a home in the city's completely white suburbs.[41]

The LIA denied that lead's harms were as dire as public health researchers had painted them, and they successfully resisted the passage of legislation to limit lead exposure. This is because, as *Lead Wars* describes, the LIA was allowed to set its own limits for lead exposure. It misleadingly labeled these "standards," as if they had been imposed by public health or legislative agencies intent on protecting children's bodies and brains. In fact, they were created and adopted by an industry intent on minimizing its losses, and these "standards" reflected the interests of the industry, not of children's health.

In the absence of legislation mandating lead containment or removal, landlords resisted paying for lead abatement, and some municipalities did not make them do so. Even in cities like Baltimore, where laws prohibited the renting of tainted homes, some landlords rented the homes anyway, or found it cheaper to abandon the properties, lead and all, as the city looked the other way and declined to prosecute them.

Even today, tainted houses in Baltimore and other cities are sometimes fobbed off on unsuspecting families who buy the homes at reasonable prices, unaware that they harbor lead. The new homeowners must then bear the costs of cleanup, abatement, and the possible lead poisoning of their family.[42]

Lead's Effects on the Body

Between 1999 and 2004, U.S. African American children were 1.6 times more likely than white children to test positive for lead. Among the poisoned, black children were three times more likely to harbor extremely high lead levels of 10 micrograms per deciliter or higher, the level of lead poisoning that entails the most damaging health ills.[43] And lead is cumulative, building up in the poisoned person over time.

Lead poisoning affects adults, who commonly suffer stomach, kidney, brain, and nervous system injury, as well as high blood pressure and behavioral problems. But children exposed to lead face worse consequences at the same doses. Lead harms nearly every bodily organ, and can slow the child's growth. It can also cause hearing loss, headaches, weakness, muscle problems, memory loss, and trouble learning and thinking clearly. Angry, moody, or hyperactive behavior and other personality changes are common.

African American children's greater lead exposure serves to depress their average IQ. In the 1980s, 35 percent of African American children who lived in cities had high blood levels of 10 µg/dL or higher, but only 5 percent of white children did.[44]

During this period, when average U.S. children's blood lead levels hovered between 12 and 17 µg/dL,[45] those suffering lead poisoning severe enough for admission to the hospital where I once worked often presented with acute poisoning. A fight to save the child's life ensued, and coma and dramatic symptoms were common enough that doctors, out of necessity, were usually focused on keeping the child alive, not saving her intellect.

As I detail at length in Chapter 4, newborns and very young children suffer lead exposure even from food and water. But exposure escalates in toddlerhood, when a child is able to move about on his own, touching lead chips, and mouthing hands and objects contaminated with lead dust. Children who swallow lead absorb half of it; their gastrointestinal systems are larger relative to their body size than those of adults; as a result they absorb *five hundred times* more lead per exposure than adults.

To understand how lead harms the brains of children, one must first understand that their developing brains feature an exquisite sensitivity that magnifies the effects of very small exposures that would not harm an adult or even an older child. Many poisons, including lead, injure children in ways that vary according to their stage of development.

For now it's just important to note that as the cells of the fetal brain develop, these neurons not only grow, but also blossom into a preordained complexity guided by an intricate, precise choreography. The choreographers are neurotransmitters, including endocrine hormones, which closely control when each cell will undergo differentiation and migrate to an exact locus within the brain.

Lead sabotages this precision, with portentous results: Brain cells may grow to the wrong size or in an incorrect configuration. They may migrate to the wrong portions of the brain, or fail to migrate at all. As a result, essential neurological structures and connections may become misshapen or may never appear at all, leading to brain damage that is irreversible.[46,47] This damage continues long after birth, and its full import may not appear until the twenties, when the brain matures. Or even later: the corpus callosum that unites the two hemispheres of the brain, for example, is a late bloomer, often completing its maturation in a person's thirties.

Moreover, this derangement of the brain's growth and architecture doesn't require acute, high-dose exposure; it is triggered by chronic, low-level lead exposure that may never be recognized, and may have no discernable symptoms for years, or even decades.[48] This is precisely the sort of exposure that preferentially haunts the nation's children of color.

Lead poisoning cannot be cured, but good nutrition, and in extreme cases, chelation—an inpatient procedure where children are given substances that bind excess lead so the body can excrete it—can lower lead levels. Chelation is not a cure-all, however, because it sometimes must be repeated, and sometimes it worsens the poisoning.

The damage is not confined to disorganization and malformation of the brain. Early exposure to lead also produces epigenetic changes that can reprogram genes, changing their expression in a manner that further heightens the risks of disability and a variety of disorders ranging from heart disease to colorectal cancer triggered by lead-induced DNA damage. The result may not become apparent until stress on the immune system later in adult life triggers failure.[49] And since such reprogrammed genes can be passed from generation to generation, this harms not only the poisoned child but also his own children, an example of the complex interplay between genetics and environmental exposures.[50]

When this genetic transmission causes brain damage leading to lowered faculties, it occurs as a result not of innate inferiority, but of a chemical insult.

Moreover, a 2007 report in *Neurotoxicology* revealed that genetic damage caused by lead also sensitizes children to later neurological assaults, weakening the poisoned person's ability to resist subsequent environmental injury.[51]

Unfortunately, the mind-destroying consequences of low-level lead exposure went unrecognized for many decades. In fact, lead poisoning was initially thought to harm children only when their blood level reached 15 micrograms per deciliter of blood (15 μg/dL). Later studies showed that children were also being poisoned at lower levels, so the toxic level was revised downward. Unfortunately, health policies in cities like Baltimore mandated no treatment for children whose blood-lead levels were less than 20 μg/dL, even though the threshold poisoning level was set at 10 μg/dL and experts argued for 5 μg/dL. Today, the CDC states that there is no threshold and no safe level of lead exposure.

Behavioral Fallout

The vanishingly low levels of lead that catalyze brain damage cause aggressive, inappropriate, and sometimes criminal behavior. Researchers blame pollutants like lead for a rise in disorders like ADHD, conduct disorder, and autism, which helps to explain why only 56.4 percent of lead-exposed Baltimore students graduate from high school. (The national rate is about 80 percent.)[52]

To better quantify the poisonous metal's role in behavioral problems, the director of Fordham University's Neuroscience and Law Center, Deborah W. Denno, conducted longitudinal studies in which she analyzed hundreds of biosocial factors that correlated with an increased likelihood of violent crime among 1,000 youths. After following the behavior patterns (compiled from such sources as their parents and school records) of 301 boys in the Pittsburgh school system, Denno measured their bone lead. After correcting for race, education, and neighborhood crime rates, the highest lead levels were found among boys who engaged in more bullying, shoplifting, and vandalism. Except for those boys with the lowest levels of lead in their blood, their behavior had worsened as they aged.

In another study, Denno followed 487 young African American men in Philadelphia from birth to age twenty-five. They all shared the same urban environment and school system. She assessed three hundred variables including blood lead, and found that childhood blood lead was the single most predictive factor for disciplinary problems and juvenile crime. It was also the fourth largest predictor of adult crime.[53]

Others also blame lead for rising U.S. crime rates. In 2007, Amherst economics professor Jessica Wolpaw Reyes released her analysis showing that the reduction in gasoline lead was responsible for most of the decline in U.S. violent crime during the 1990s.[54]

In his 2016 *Mother Jones* article "Lead: America's Real Criminal Element," political blogger Kevin Drum details the evidence supporting lead as a driver of national crime rates.[55] Like James Q. Wilson and Kevin Nevin, he correlates the rise and fall of crime rates with the addition and elimination of lead from gasoline. In broad strokes, as lead was banned from gasoline, crime fell, and it rose along with lead's increasing presence in interior paint and toys. This is persuasive, although it doesn't rise to the

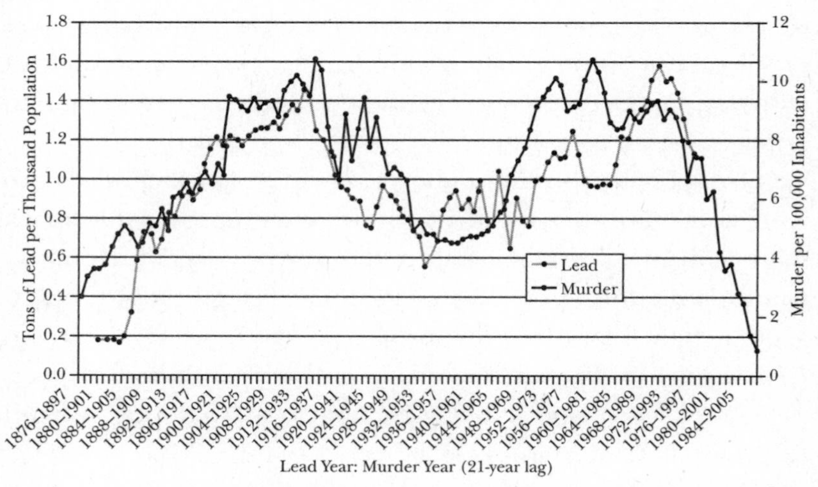

Murder rates climb with rising lead-exposure concentrations, and dip as blood lead falls.

power of absolute proof: there are flaws in the data interpretation, many of which Drum himself discusses elsewhere.

Criminal Element?

However, one methodological flaw that remains below the radar is the folly of equating arrest and incarceration with crime, especially when people of color are concerned. Being arrested or even convicted does not mean that a person has committed a crime. Neither is arrest a racially equitable response to behavior. For example, 92 percent of people arrested for marijuana possession in Baltimore in 2010 were African Americans, but this is because black Baltimoreans are more than *5.6 times* more likely to be *arrested* for possession of marijuana than whites, even though marijuana use among the races is similar.[56] Moreover, false arrests occur: In 2017, prosecutors dismissed thirty-four criminal cases after body-cam footage showed Baltimore police officers planting drugs at crime scenes.[57]

Especially for African American boys and men, whose behaviors are more strictly scrutinized and judged than those of whites, a wide spectrum of social frictions contributes to a greater chance of interaction with the legal system and can lead to arrest, injury, incarceration, and even death—even when no crime has been committed.

Several recent studies indicate that even black children are less likely than white children to be viewed as innocent. Black schoolchildren have been assaulted and arrested for politely contradicting teachers' statements, for wearing braids, for playground fights, and for "inappropriate" clothing. One African American girl was arrested when her properly conducted, teacher-supervised chemistry experiment went awry, causing a small explosion in the classroom lab.

U.S. jails hold hundreds of thousands of people who are there not because they are guilty but because they are too poor to make bail or to commission the tests, including DNA tests, that would prove their innocence. Since 1989, DNA testing prior to conviction has proven that tens of thousands of prime suspects were wrongly accused, wrongly identified, and wrongly pursued.[58] Stricter "stop and frisk" scrutiny of African American and Hispanic neighborhoods, racial disparities in the ability to make accurate facial identifications, errors, intentional or unintentional, by law enforcement, and the perception of black Americans as criminals all feed unjust incarceration.

Most people are aware that the Innocence Project and its offshoots have freed hundreds of innocent men from death row, most of them African American. Between 2000 and 2008, for example, between 50 and 70 percent of the incarcerated men exonerated by DNA technology were black or Hispanic.

Most of the convictions disproved by DNA evidence involve African American men wrongfully convicted of assaulting white women. "This is a crime that seems associated with many false convictions," said Peter Neufeld of the Innocence Project in 2001.[59]

But most prisoners are not facing death sentences, nor do they have skilled Innocence Project lawyers to reevaluate their cases.

As a result, determined University of Michigan law professor Samuel R. Gross, tens of thousands of innocent people are trapped in jail: "If we reviewed [all] prison sentences with the same level of care that we devote to death sentences, there would have been more than 28,500 non-death-row exonerations in the past fifteen years rather than the 255 that have in fact occurred."[60]

Finally, crime is multifactorial, and so are its putative chemical risk factors, including environmental poisons like mercury, manganese, and pesticides, all of which cause behavioral fallout, including impulsivity and criminality.

Slow Realization

Why did it take so long for scientists to realize that low levels of lead were causing rampant intellectual deterioration and dramatic behavioral problems? In 1982 David P. Rall, the former director of the U.S. National Institute of Environmental Health Sciences, wrote, "If thalidomide had caused a ten-point loss of IQ instead of obvious birth defects of the limbs, it would probably still be on the market."[61]

In fact, thalidomide and its analogues have long been back on the market,[62] but this fact doesn't weaken Rall's observation that we have been slow to recognize the invisible cognitive fallout from lead—and we have been even slower to act.

The late child psychiatrist and pediatric researcher Herbert L. Needleman devoted his career to removing the scourge of childhood lead poisoning. The son of a furniture salesman and a homemaker, he graduated in 1952 from the University of Pennsylvania medical school. After a short stint in the Army, he completed his medical training at a North Philadelphia community health clinic, where he spent his days treating children

Pediatrician Herbert Needleman spent a career documenting and combating insidious lead poisoning in urban U.S. children. (Courtesy of Jim Harrison)

whose brains had been damaged and intellects short-circuited by exposure to high levels of lead.

His office window opened into a nearby playground where seemingly well children played every day. They had not been exposed to high doses of lead, but they lived in old, crumbling housing that harbored lead paint, so they were bombarded by low doses every day. Needleman wondered, was there an effect from these low, constant exposures to lead? He struggled to devise a viable way to find out—testing hair, blood, or fingernails was not accurate enough to register low doses. Bone biopsies were accurate but painful: taking them just to test a theory is unethical.

One day, Needleman gave a talk at a local African American church and afterward was approached by a young boy who clearly admired the doctor and sought to share his own ambitions. But as he listened, Needleman grew dismayed. "He was a very nice kid, but he was obviously brain damaged. He had trouble with words, with prepositions and ideas. I thought 'how many of these kids who are coming to the clinic are in fact a missed case of lead poisoning?' "[63] Needleman realized he was seeing the same signs of brain damage that his profoundly lead-poisoned patients displayed. Now it was more than an academic question: he had to discover whether low-level lead exposure was also destroying children's mental capacities.

Tooth Fairy, M.D.

But how to test this? As noted above, hair, skin, and nail samples were not sensitive enough, and bone biopsies were accurate but painful, and therefore unethical just to satisfy curiosity. Needleman eventually hit upon a painless, accurate method: "Herb became the Tooth Fairy," Dr. Bernard Goldstein recalled for the *New York Times*. Needleman realized he could test the baby teeth of affected children, and he began paying children in the city's poor neighborhoods with disproportionately African American populations for each baby tooth that fell out. Needleman found that lead was cumulative, and that the average child had lead levels five times higher than those of his suburban peers. And an IQ gap of four points separated the children with the highest exposures from the lowest. These children who had not been suspected of being harmed by lead had relatively low IQ scores, poorer facility with language, and shorter attention spans.[64] When teachers rated the exposed children's classroom performance, a host of behavioral issues emerged, from attention deficits to behavior problems. Ten years later a correlation persisted between their childhood lead levels and reading delays.

Needleman was also acutely aware that he was fighting a racial scourge. When asked why he thought the government was reluctant to spend the funds necessary to completely eliminate lead from housing, he replied, "Well to begin with, it's a black problem."

The increasing focus on "personal responsibility" replaced accountability demands on the lead industry. Rather than hold manufacturers liable, the onus of avoiding harms increasingly fell on the victims, a cultural shift with profound legal implications. Cities sometimes failed to sue landlords who did not eliminate lead from their properties, worrying that if landlords were pressured to comply, they would simply abandon their properties, robbing a city of its tax base.

Instead, parents of poisoned children were told that it was their responsibility to keep their homes clean and free of lead dust and paint chips. Public health experts suggested abatement regimens that included training parents in the use of cleaning products like Spic & Span, implying that parents were failing to clean frequently or properly. Health workers also suggested that toddlers be confined to playpens to prevent contact with lead, a laughably unworkable—and unhealthy—strategy.

Full abatement of lead from the home was the only solution known to protect children, but its cost was rejected as prohibitive.

Needleman disagreed with this fobbing off of responsibility onto the injured, and he refused to accept that intervention was futile and too costly.

Moreover, he saw lead poisoning not as a discrete medical issue but as a symptom of social ills: medical and economic disparities that arose from racial discrimination.

And as Rosner and Markowitz describe in *Lead Wars,* he devised a visionary plan to address lead poisoning in this context. Under his plan, lead would be completely abated, and the work of properly sanding, repainting, and removing and replacing lead-tainted housing materials would be performed by unemployed workers under a federal program that resembled a latter-day Work Projects Administration. If followed, Needleman argued, his plan would remove the lead hazard while simultaneously addressing rampant unemployment in the lead-poisoned communities.[65]

In 1989, the plan's price tag of $10 *billion* was rejected as far too expensive, although Needleman noted at the time that no one decried Congress's plan to spend $11.6 billion to build new prisons.[66] Since then, our government has spent far more than that amount dealing piecemeal with some of the aftereffects of lead poisoning, and that doesn't include the cost of defending lawsuits against property owners and the unethical, shortsighted plans of institutions like the KKI. Even so, lead has still not been eradicated, and two of every three poisoned children in Baltimore are living in the same pre-1950 rental homes that Needleman's plan sought to abate.

But Needleman's findings were a key component in the 1974–1978 bans on lead in gas and paint, and his participation in various legal cases on behalf of poisoning victims made him an industry target. Accusations against him culminated in scientific misconduct charges brought before the federal Office for Scientific Integrity, and later, before his own institution, the University of Pittsburgh.

He was exonerated in both cases.

In a 2005 interview, Rosner and Markowitz wrote, "Dr. Needleman was asked whether the attack on his credibility was meant to scare off other researchers looking into environmental toxins. 'If this is what happens to me, what is going to happen to someone who doesn't have tenure?' he replied."[67]

In the past two decades, Maryland has passed a stronger law requiring landlords to cover or remove lead-based paint that's peeling, chipping, or flaking. But properties are rarely monitored for compliance. And although the number of cases has fallen, at least 4,900 children were diagnosed with lead poisoning in the decade between 2005 and 2015 in Maryland alone. In reality, the number is probably much higher, because approximately 7 million lead poisoning tests processed nationwide by Magellan's LeadCare Testing Systems underestimate lead levels, giving erroneous results.[68]

"If rich white kids were getting poisoned, there would be a law on the books that says 'No lead in houses,'" lawyer Brian Brown told the *Baltimore Sun* in 2015.[69] Today, Needleman's $10 billion price tag looks like a bargain, even in adjusted dollars.

Blame the Victim 2.0

Young African American men are killed by police officers at nine times the rate of other Americans—1,134 were killed by police officers in 2015 alone.[70] There is no salient justification for the disparity: black men are no more likely to be armed or aggressive than their white peers.

Police brutality is another risk to life and health in blighted communities of color. At least two of the noncombative, unarmed African Americans killed by police as they went about their daily routines were victims of something besides police violence: both Freddie Gray, twenty-five, and Korryn Gaines, twenty-three, suffered from lead poisoning.

In the 1990s, Gray lived from the age of two in a row house in Baltimore's blighted, lead-soaked Sandtown-Winchester area, according to a 2008 lawsuit filed by Gray and his siblings against the property's landlord. Average life expectancy in Sandtown is lower than in North Korea, and lead may be a factor. In his neighborhood and in several nearby census tracts, between 25 and 40 percent of children—that's two out of every five—tested between 2005 and 2015 had elevated lead levels.[71]

As a result of being exposed to lead for two decades, Freddie Gray suffered developmental problems. The Gray family's case was settled for an undisclosed amount.

In some newspaper accounts of his killing, I was surprised to read vague speculation that these lead-poisoning victims might have helped bring about their own deaths because of unspecified "behavioral problems" that are often linked to lead exposure. No such behaviors had been documented in them during the events, and this speculative claim recalls the blame-the-victim indictments of parents by lead-based industries.

Quite aside from the questions of guilt and innocence in deadly police encounters with peaceable, unarmed African Americans, we mustn't use speculative medical scrutiny to demonize some victims of lead poisoning while winking at others. This bias compounds the victimhood of sufferers like Freddie Gray and Korryn Gaines by conflating their poisoning injuries with criminality.

This same animus drives some to criminalize the parents of lead-poisoning victims. In August 2015, Kenneth C. Holt, Maryland's secretary of housing, community, and development, dismissed the plight of poisoned children by speculating that their mothers were deliberately exposing their own children in fraudulent attempts to obtain better housing. A mother, he said, might place "a lead fishing weight in her child's mouth [and] then take the child in for testing." He later admitted that he was not aware of any mother actually doing what he had suggested.[72] Such demonization of the poisoning victims is not new: In their book *Deceit and Denial: The Deadly Politics of Industrial Pollution,* Gerald Markowitz and David Rosner recount how a lead industry representative referred to the lead-poisoned children of Baltimore as "little rats" and their mothers as "overfecund imbeciles."[73]

This baseless indictment of black parents illustrates a readiness to blame them, as well as a default stance of callous disregard for children's health by the very institutions charged with protecting it.[74]

For all these reasons, our concerns about lead poisoning should be framed in a medical context, not a demonizing one, urges Ruth Ann Norton, who directs Green and Healthy Homes Initiative, a twenty-two-state agency dedicated to environmental health. "I don't know what Freddie Gray did between the ages of three and twenty-five, but if he had been able to read well, had gone to school... [if] his family wasn't just fleeing from one house to another, the likelihood of him not being on that corner would have been a whole lot better. We know that.... When do we want to stop dumbing down our kids?"[75]

At least 37,500 Baltimore children, nearly all of them black, were diagnosed with lead poisoning between 1993 and 2015, like Gray and his sisters. But only one in five black children is now tested.

In West Baltimore, Olivia Griffin's children were raised in lead-tainted housing that slipped through the cracks of municipal lead monitoring. But she has now completed a job training program and found untainted housing through the Green and Healthy Homes Initiative. Still, her six-year-old son Nazir is paying the cognitive price for growing up in a lead-laden atmosphere. He "acts out a lot" and was slow learning to talk, she said. She took him to a speech therapist several years ago, but his speech still gets garbled sometimes. "You just have to be around him for a while so you can understand it."[76]

The lead that once permeated the entire nation's homes and air now persists in enclaves of color. These include Flint, Michigan, which garnered the nation's attention when its people finally learned that their neighborhoods had been secretly flooded with poisoned water; East Chicago; and New York City, whose poisoned denizens still suffer in quiet desperation as a result of segregation, deception, and fatal greed; among many others.

Troubled Waters: Poisoning Flint

When water is pure, the people's hearts are at peace.

— Kuan Tzu

In 2014, Joe Clements, a seventy-nine-year-old who had been raised by adoptive parents, turned to DNA testing to seek out lost branches of his family tree. He'd lost many loved ones over the years, and he was eager to connect with his biological family. In the autumn his search bore fruit in the form of Randye Bullock, a half-sister, along with a whole new branch of his family.

The joy in their reunion was mutual, and Clements decided to move nearer to them. That fall he had begun life with his newfound family in a new city—in Flint, Michigan.

But just a few months after he arrived, aging but active and healthy, he began to sicken. He rapidly lost weight and suffered constant stomach pain and gastrointestinal distress. Fatigue washed over him while climbing steps or with the slightest exertion, and soon he lacked the energy to go out at all.

But Clements lived with more than pain and lethargy. His short-term memory deserted him, leaving him confused, emotionally brittle, and constantly irritable. He found himself unable to follow through on the simplest tasks. He couldn't sleep, his hands shook, and sometimes, his legs did too. "I'm always exhausted and now I'm having lung congestion and memory loss," he said. "I don't even know the extent of damage to my body."

Clements believed in drinking lots of water to keep his kidneys functioning well, so after he noticed his body weakening, he quaffed tap water even more liberally than he had before. And why not? City and state officials had repeatedly assured Flint families that their water was safe.[77]

However, younger members of his family were beginning to sicken as well, and unlike Clements, they had lived in Flint long enough to notice changes in their water. In the spring of 2014, yellowish water that smelled heavily of bleach had begun gushing from spigots.

The worst contaminant could not be seen or tasted: lead. Flint water's lead levels were so high that it fell into the EPA's classification for hazardous waste.

The service lines in Clements's neighborhood were also made of lead. And because the water wasn't infused with the necessary anti-corrosive chemicals, lead, iron, and chlorine leached from these aged, encrusted lead pipes into the already toxic tap water. By early 2015, its lead level was nineteen times higher than before Clements had moved to town.[78] Moreover, despite the assurances to the people of Flint, General Motors was so worried about the corrosive water that it stopped using it at its engine plant out of fear of damaging its equipment.

Less than two years after he arrived in Flint, Joe was diagnosed with a rare, aggressive kidney cancer. A year after that, he was dead.

Lead exposure is suspected not only in Clements's mental deterioration, but also in his kidney disease, says Michigan State University nurse educator Patrick Hawkins, Ph.D. "Even before the lead crisis, Flint residents had 2.5 times the national rate of kidney disease. No one knows precisely why. But I fear it will go much higher as we learn more about lead exposure. Lead moves out of the bloodstream and attacks organs."

America's lead crisis is a national threat to American bodies and minds. And as most now know, residents of the heavily African American city of Flint have been forced to drink lead-tainted water since 2014, the result of bureaucratic shortsightedness and deceit.

But today's scandal is not Flint's first brush with toxic notoriety. For at least eighty years, children growing up in the bustling industrial city center were surrounded by heavy metals that caused neurological disorders.

Then, as Michael Moore chronicled in his 1989 documentary *Roger and Me,* General Motors cars were displaced by more popular models from Volvo, Honda, and Toyota, and GM abandoned Flint, taking most of its tax base with it. In its wake, Flint was left an impoverished, poisoned city.

By early 2014, nearly half of Flint's 100,000 residents lived below the poverty level. Forty-two percent still do, giving Flint one of the highest poverty rates in the nation for a city its size. In an effort to address the $9 million deficit facing Flint's water supply fund, Governor Rick Snyder instituted an emergency management system, switching the city's water supply from Lake Huron (via the Detroit Water and Sewer Department) to the Flint River, which was tainted from decades of use as an industrial dumping site for GM and the other industries that once lined its banks. Snyder rejected supplying the necessary corrosion protection for Flint's lead pipes as too expensive.

Now, the city is enveloped in a public health emergency, with high levels of lead in its water supply and in the blood of its children. Although lead poisoning is devastating to adults like Joe Clements, it is much more deadly to children—with far smaller amounts causing IQ loss, hearing loss, attention deficit hyperactivity disorder, dyslexia, and even death.[79]

And as usual, race is a more powerful determinant of environmental exposure than socioeconomics.

Flint residents, news media, and health care professionals demanded that their discolored, foul-tasting water be tested and corrected, but nothing was done. Although the Flint water crisis is now acknowledged as racially disparate lead poisoning, it was originally described as a problem that plagued a "lower socioeconomic group." And even after the racial disparity was documented, references to "socioeconomic" exposure persisted—sometimes even among those who elsewhere acknowledge its racial nature. For example, in late 2015, Flint pediatrician Mona Hanna-Attisha, M.D., became a medical hero when she first brought national attention to the lead poisoning in the *American Journal of Public Health.* She described the city's blood-lead elevations as exclusively affecting "socioeconomically disadvantaged neighborhoods.... As in many urban areas with high levels of socioeconomic disadvantage and minority populations," she wrote, "we found a preexisting disparity in lead poisoning."[80]

Flint is a majority-minority city, meaning that more members of racial minority groups than whites live there. Most—65 percent—of its 99,000 residents are African American and Latino. Forty-two percent are poor.

"Socioeconomic status" (SES) is a variously defined term that refers to some interaction of social and economic factors—rather than to race. When a problem is ascribed to SES, many assume that the issue is driven by poverty and social conditions, such as lack of access to education, rather than by race or racial bias. As a result, socioeconomics is often set in opposition to racial bias, suggesting that health issues such as lead exposure are driven by economics or nonracial social issues rather than racial issues. In fact, socioeconomics and race are inextricably intertwined.

Residential segregation by race, an example of an institutional racism, has created racial differences in education and employment opportunities, which in turn produce racial differences in SES. In addition, segregation is a major determinant of racial differences in neighborhood quality and living conditions including access to medical care.[81]

Race also influences the "economics" component of socioeconomics: Any issue driven by poverty will necessarily have a disparate effect on marginalized ethnic and racial minority groups. Thirty-nine percent of African Americans are poor, as are 43 percent of Hispanics. Twenty-eight and a half percent of Native Americans and Alaska Natives live in poverty, nearly twice the rate of non-Hispanic whites.[82] Furthermore, African Americans and Hispanics earn fifty-nine and seventy cents, respectively, for every dollar of income that whites receive. The racial differences in wealth, that is, in a household's economic assets and reserves, are even more stark: for every dollar of white wealth, Asian households have eighty-three cents, Hispanics have seven cents, and black Americans have six cents.[83]

Because race is actually a component of the most pertinent definitions of socioeconomics, putting socioeconomics in rhetorical contrast to race is illogical. Instead, we must understand that race is an important component of SES.

Moreover, scientific reports have consistently demonstrated that race poses a stronger risk factor for the placement of environmental poisons than poverty. Poverty is a driver of environmental exposures, but race is a greater driver. Reflecting this truth, the definition of SES in public health spheres explicitly includes race as a factor. As Professor David R. Williams of the Harvard T. H. Chan School of Public Health notes: "All indicators of SES are strongly patterned by race."[84]

Accordingly, although initial media accounts didn't mention race but referred to Flint's poverty as the key risk factor, with time, accounts acknowledged that race was the salient vulnerability.

"It Is Just a Few IQ Points..."

Some think racial bias encouraged the callousness with which even professionals contemplated the poisoning of children in Flint. For example, 274 pages of e-mails retrieved from the office of Governor Snyder reveal a dismissive stance toward the people, including pediatricians, voicing their concerns: they were derided as an "anti-everything group."

In a February 2015 e-mail, a state nurse dismisses a mother's concern about her son's brain damage due to his elevated blood lead level, writing, "It is just a few IQ points.... It is not the end of the world."[85]

This indifference contrasts starkly with a 2013 Detroit Department of Health and Wellness Promotion study of the city's schoolchildren that linked early childhood lead exposure to educational testing data from the Detroit public schools. Researchers found lower academic achievement in mathematics, science, and reading among elementary and junior high school students who had suffered lead exposure.[86]

But polluting industries also downplay the significance of "a little" IQ loss, says Bruce Lanphear, an environmental health expert at British Columbia's Simon Fraser University. "The chemical industry argues that the effect of toxins on children is subtle and of little consequence. But that is misleading."[87]

To be sure, not every chemical is harmful, or harmful in small doses. But lead is: You'll recall that the CDC states that there is no safe level of exposure.

Science tells us that even a small IQ drop could indeed be the end of, if not a child's world, certainly his dreams. This is clear once one understands what the loss of a few IQ points betokens to a child and to our society.

Scientists find it difficult to determine the number of lost IQ points due to lead exposure because there are so many variables. These include: the type and duration of exposure; the developmental stage at which exposure began; the amount and sort of co-exposures a child is subject to—mercury, for example; and whether the child has benefited from compensatory enrichment like prekindergarten classes or treatments like lead chelation.

But the best, strictly regulated analyses consistently show that children forfeit approximately one-quarter to one-half of an IQ point for every µg/dL of blood lead.[88] This means that a child with a blood lead reading of 10 µg/dL (for a long time the lowest threshold for diagnosed poisoning) loses about five IQ points.

An IQ drop of five points makes a child a bit slower to learn, and reduces her memory capacity as well as her ability to read and calculate. Her tests will reveal consistently lower scores. But in addition to IQ loss, lead poisoning sabotages a child's ability to learn by engendering learning disabilities, hearing loss, balance disorders, hyperactivity, perceptual disorders, attention deficit disorders, and a reduction in perceptual reasoning skills. Lead also causes violent tendencies, greater impulsivity, and disruptive classroom behavior. This synergy renders the child less able to compete in the classroom and more likely to drop out. The lead-poisoning damage is also cumulative, and affected adults are less able to compete professionally and to hold employment, making them more likely to be unemployed or underemployed.

"Lead exerts a downward pull on an individual's cognitive abilities over time regardless of where they start out in life," wrote Aaron Reuben of Duke University in Durham, North Carolina. As adults, these children will earn less than their unaffected compatriots—on average $90,000 less—while working at lower-status jobs than their parents. A wealth of studies document that the stress of racism also exacerbates the damage wrought by the IQ loss.

IQ loss has national consequences as well. Even a small

average IQ loss skews the distribution of mentally retarded and intellectually gifted children in an ominous manner.[89] Using CDC data, Lanphear calculates that *an average five-point drop in IQ among U.S. children would result in 3.4 million more children being intellectually disabled or mentally retarded.* Simultaneously, such a seemingly small drop would reduce the number of "intellectually gifted" children by half, and double the number of children that meet the criteria for "intellectually impaired."[90] With this in mind, consider the stunning results of a study conducted by Harvard's David Bellinger, which shows that, as noted in the Introduction, lead exposure has cost us a total of *23 million lost IQ points* nationwide every year—even more than the 13 million IQ points lost to pesticide exposure as documented by a European Union study.[91]

Flint remains embroiled in a public health emergency, with high levels of lead in its water supply and in the blood of its children. But like the victims in Baltimore before them, Flint's victims have been demonized.

Flint residents are charged some of the nation's highest rates and taxes for their poisoned water. And when they cannot pay, the city takes draconian legal action, sometimes forcing residents from their homes. The Genesee County Land Bank, Flint's

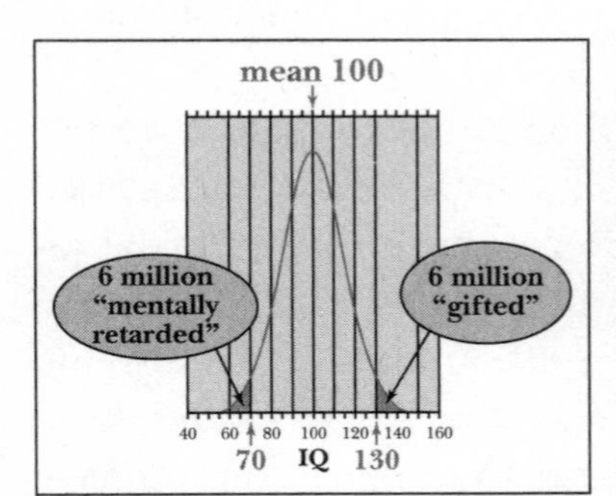

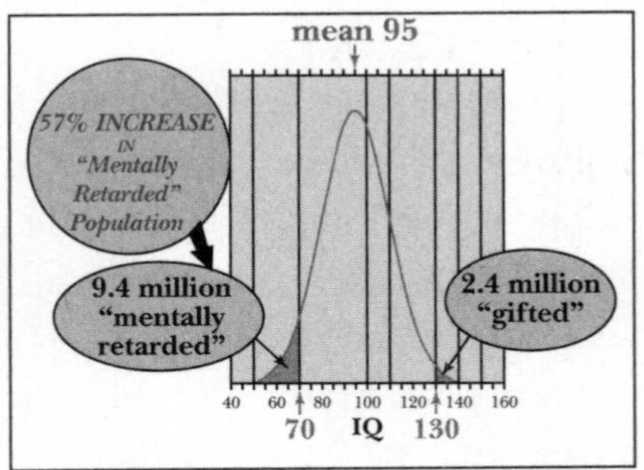

An average 5-point IQ loss causes the number of "mentally retarded" people in a population to balloon by 57 percent, and causes the number of "gifted" to fall precipitously.

largest property owner, takes over tax-foreclosed properties, demolishes them, and orchestrates rehabilitations and sales. It is blamed by many for driving longtime Flint residents from their homes. When a journalist asked its sales manager, Phil Stair, about Flint's tainted water, he responded, "Well, Flint has the same problems as Detroit, fucking *niggers don't pay their bills;* believe me, I deal with them." This is nonsense: a state commission found that "systemic racism," not African American "deadbeats," as Stair also referred to them, caused Flint's disaster. Amid the resulting outrage, Stair resigned.

The price tag for replacing Flint's corroded lead pipes is now estimated at $1.5 billion,[92] which Flint doesn't have. The Environmental Protection Agency has offered no timetable for replacing the pipes and providing potable water, so the foreseeable future portends a city of children with lowered IQs and a community robbed of its future potential. As whistleblowing Detroit pediatrician Mona Hanna-Attisha, M.D., told the *New York Times,* "If you were going to put something in a population to keep them down for generations to come, it would be lead."[93]

But Rosner and Markowitz point out a graver problem that extends far beyond Flint: "The mix of racism and corporate greed that have put lead and other pollutants into millions of homes in the United States. The scores of endangered kids in Flint are just the tip of a vast, toxic iceberg."[94]

This iceberg encompasses the nation. Ten miles away, in Detroit, thousands of Motor City homes still bear lead paint and more than fifteen hundred children were lead poisoned in 2014, but that city only had enough money for one hundred to two hundred lead paint abatements annually. In Detroit, where the population is 84 percent black and housing is notoriously dilapidated and abandoned, 80,000 of the city's 380,000 properties are considered blighted, according to the *New York Times.* This creates a risky situation for children living in those homes, as well as in neighboring properties. Eight percent of children who were tested for lead poisoning in Detroit had elevated blood-lead levels in 2013 (≥5 micrograms per deciliter), sixteen times the national average reported by the CDC and state health departments.

In a national pattern, African American children are nearly three times more likely than white ones to suffer the most highly elevated, most damaging blood-lead levels. Majority-minority cities have the highest number of children with these elevated levels, although African Americans constitute only about 13 percent of the U.S. population. In Savannah, Georgia, which is 57 percent black, 5 percent of children had elevated blood-lead levels, ten times the national average. Birmingham and Montgomery, Alabama, both of which are 50 percent African American, have similarly high rates of lead poisoning: 3 and 4 percent of all children, respectively.

The water supplies of some Native American communities have been far more heavily poisoned than Flint's, for a much longer time. For example, uranium mine and mill contamination renders the water of Navajo communities radioactive in New Mexico and other western states. In 1979, Justin Garner wrote, "A spill north of Churchrock, New Mexico left an immense amount of radioactive contamination that down-streamers, today, are currently receiving in their drinking water. A mostly Navajo community in Sanders, Arizona has been exposed to twice the legal limit allowable for uranium through their tap."[95]

The aging water-delivery infrastructures of hundreds of U.S. cities, including Philadelphia and Chicago, feature lead pipes that now leach the poisonous metal, thanks to the industry's success in blocking proposed regulation in the 1920s. Throughout the nation, once-pristine water sources are befouled by a poisonous brew of raw and "treated" sewage, industrial chemicals, animal offal and excrement, and pharmaceuticals.[96]

Although lead is the primary concern in cities like Baltimore, Flint, and East Chicago, researchers also worry that other chemicals, like total trihalomethanes, or TTHMs, and the extra chlorine that Flint pumped into its system to combat bacterial growth, are linked to sickness and higher miscarriage rates.

In October 2011, in New York City, a small one-bedroom apartment in a six-story building near Fordham University and just two miles from where I lived seemed to be just what Zaimah Abdul-Majeed and her husband were looking for. At about $1,000 a month, it was that Holy Grail of NYC real estate—a clean, affordable, apartment—even if that apartment was in the Bronx and had to house four people: the couple and their twin one-year-olds.

But the freshly whitewashed walls could not hide its downside for long. Two months after they moved in, the Abdul-Majeeds had a rude awakening in the form of a call from their pediatrician's office: their daughter Zoe's blood lead level was 21 µg/dL—more than four times the amount deemed by the CDC to predispose her to lowered intelligence and a heightened risk of behavioral disorders. "I was shocked," said Ms. Abdul-Majeed. "She's starting off with a delay in her life, and it clearly wasn't her fault." Since 2004, New York City's Local Law 1 has held landlords accountable for lead contamination. As a result the rate of poisoning has fallen by 66 percent. But the city's worst scofflaw landlords are responsible for hundreds of violations. According to the *Huffington Post,* the Abdul-Majeeds' landlord was one of the top ten in lead poison violations. These ten landlords alone were cited more than 1,000 times between November 2013 and January 2016.

The city's Department of Housing Preservation and Development (HPD) issued more than 10,000 violations for dangerous lead paint conditions in apartments where young children live. Half were in just 10 percent of the city's zip codes—low-income neighborhoods of the Bronx, Brooklyn, Harlem, and northern Manhattan. About 8 percent had lead levels exceeding 15 parts per billion. And 83 percent of the schools in these minority areas had at least one water spigot with a lead level above the legal exposure threshold.

In New York City, federal prosecutors recently opened an investigation into lead hazards found in the city's public housing.[97]

New York City breaks lead poisoning rates down by race in its reports, but its jurisdictions aren't required to report that information to the Centers for Disease Control and Prevention. This means that no data sets exist to distinguish the differences in elevated blood-lead levels among races. Moreover, many counties and municipalities don't collect or report lead poisoning data to the CDC at all.

The Ghosts of Lead Smelters Past

A 2001 *American Journal of Public Health* article revealed that when it comes to missing data on lead, perhaps the most ominous cases are those of the four hundred vanished lead smelters that operated nationwide before the creation of the Environmental Protection Agency and its regulations in 1970. In 2012, environmental scientist William Ecke revealed their past existence, but we can't know how much lead they leached into the environment, or how much persists. *USA Today* reporters tested soil at four hundred sites in thirteen states and found persisting lead levels that varied but were as high as ten times the EPA hazard threshold. Although the EPA was notified, it did little to test or to warn current residents that they reside on the sites of former lead smelters.[98]

The Philadelphia EPA office responded: "PA does not notify residents of potential contamination based solely on the possibility that past industrial activities may have occurred. This type of approach would unnecessarily alarm residents and community members."[99]

In the United States, 1.8 million children live in homes with dangerously deteriorating lead paint. Fifty-two percent of all residential housing units have lead paint at concentrations that may cause adverse effects in children, such as alterations in heme (blood protein) synthesis and neuropsychological deficits including decreased IQ, behavioral changes, and impaired school performance.* To learn what you can do to protect your family and community see Chapter 6, "Taking the Cure."

* Specifically, concentrations of 0.7 mg/cm^2 or higher.

CHAPTER 3

Poisoned World: The Racial Gradient of Environmental Neurotoxins

I grew up an Army brat, which made moving to a new military base, a new town, and sometimes a new country every few years the norm. But I was anchored by the adopted hometowns of Croton-on-Hudson, New York, where my maternal grandparents lived, and, thirty miles away, the Harlem home of an aunt and uncle.

Whenever I visited Croton, I'd join playmates and cousins in daily routines of outdoor games like hide-and-go-seek, skating, and cycling as we ranged for miles across vast neighborhood gardens and the pristine surrounding woods, collecting butterflies, tadpoles, and less savory samples of nature.

But Harlem's big-city sophistication contrasted sharply with my grandparents' sleepy exurban town: flitting to museums, planetariums, art galleries, films, and shopping by subway and bus ruled my days as we walked blocks on streets choked by fuel emissions in that pre-lead-abatement era. My Harlem playmates had a familiarity with culture that eclipsed that of Croton, or for that matter, on Army bases.

Eventually I realized that they had something else I hadn't encountered in Croton: asthma. It's purely anecdotal, but I cannot recall a single Harlem playmate, including my cousins, who didn't suffer bouts of wheezing, carry an inhaler, or tell of frightening nights spent struggling to breathe while suffering from coughing, chest tightness, and narrowing of their airways.

Every child in the thirty-story building didn't have asthma, of course, but it was commonplace, sometimes culminating in a white-knuckle ride to Harlem Hospital, ten blocks down the street.

As a child, I made no connection between the plight of my asthmatic playmates and the fact that the 146th Street Depot[1] and its hydrocarbon-laden emissions were right across the street from my aunt's building, which is within a large African American residential complex. And this depot, now named the Mother Clare Hale Depot, wasn't alone: nine of New York City's ten bus depots were located in Harlem.

Asthma is still common in many African American communities partly because the oil and natural gas industries located predominately in or near them violate EPA air-quality standards for smog due to natural gas emissions. "Dirty" emissions from power plants combine with motor-vehicle pollution to form ozone smog, which triggers respiratory ailments, including asthma.

In this dense city of 8.5 million people, Harlem, the South Bronx, and other heavily industrialized ethnic neighborhoods are marked by higher asthma rates and lower vigilance over environmental pollution. This results in more than 138,000 asthma attacks among New York schoolchildren and at least 100,000 missed days of school each year.

The CDC reports that the combined national annual cost of asthma includes 10 million lost school days, 1.8 million emergency-room visits, 15 million outpatient visits, and nearly 500,000 hospitalizations, to say nothing of its $14.5 billion cost in 2000. African Americans and Hispanic Americans are three to four times more likely than whites to be hospitalized or to die from asthma.[2]

One would think these airborne risks would strike more democratically, given that we all must breathe. But do we all really breathe the same air? For that matter, do we all share the same environment?

Across the nation the befouled air hangs heaviest over communities of color. University of Minnesota researchers found that about 69 percent of Hispanic children, 68 percent of Asian American children, and 61 percent of African American children live in areas that exceed EPA ozone standards, compared with 51 percent of white children. African Americans and other people of color breathe 38 percent more polluted air than whites and are also exposed to 46 percent more nitrogen oxide than whites.[3]

More than 60,000 chemicals were registered for commercial use in the United States by the 1970s, most, as noted in Chapter 2,

Top 10 States by African American Population Living within a Half-Mile Radius of Oil and Gas Facilities (2010 Census)

State	African American Population within a Half-Mile Radius	Percent of African American Population in State within a Half-Mile Radius
Texas	337,011	10%
Ohio	291,733	19%
California	103,713	4%
Louisiana	79,810	5%
Pennsylvania	79,352	5%
Oklahoma	73,303	22%
West Virginia	13,453	17%
Arkansas	10,477	2%
Mississippi	10,448	1%
Illinois	10,227	1%
Total	**1,009,527**	**86%**

Source: http://oilandgasthreatmap.com

Top 10 Metropolitan Areas by African American Health Impacts Attributable to Ozone Caused by Natural Gas Pollution

Metropolitan Area	Asthma Attacks (per year)	Lost School Days (per year)
Dallas-Fort Worth (TX, OK)	8,059	5,896
Atlanta (GA)	7,499	5,469
Washington-Baltimore (DC, MD, VA, WV, PA)	7,216	5,269
New York-Newark (NY, NJ, CT, PA)	5,235	3,821
Houston (TX)	4,256	3,111
Chicago (IL, IN, WI)	3,777	2,760
Memphis (TN, MS, AR)	3,674	2,692
Philadelphia (PA, NJ, DE, MD)	2,887	2,104
Shreveport-Bossier City (LA)	2,536	1,871
Detroit (MI)	2,402	1,751
National African American Total	**137,688**	**100,564**

without human-safety tests. They found their way into a dizzying array of consumer products, from mattresses to computers to cookware to flame-resistant sleepwear and plastic sippy cups for babies.

Today the United States has safety data for only a fraction of the 85,000-plus commercially used chemicals (the European Union puts the global number at 145,000). Public-health studies show that exposures to these are now "ubiquitous." Among the many undertested chemicals pervading our nation are manganese, high levels of fluoride,* the pesticide chlorpyrifos, dichlorodiphenyltrichloroethane (DDT), tetrachloroethylene (TCE), and the polybrominated diphenyl ethers (PBDEs) used as flame retardants, all of which epidemiologists finger as developmental "neurotoxicants"—neurological poisons that harm the brain.

* This refers to higher-than-clinical levels of fluoride. The low levels meant to discourage tooth decay that are found in fluoride-infused waters have been tested and found safe by most studies.

These six pollutants are especially dangerous to the developing nervous systems of fetuses and very young children and are prime drivers of intelligence loss.

Even exceedingly low concentrations of some toxic chemicals can have disastrous effects on intelligence and behavior, especially if exposure occurs during early brain development. For many of these chemicals, there is no apparent threshold or safe level, even though government safety standards and testing protocols assume that there is.[4]

African Americans and other people of color are 79 percent more likely than white U.S. residents to live in neighborhoods where these potent "brain thieves," emitted from bus depots, lead smelters, petrochemical plants, refineries, garbage dumps, incinerators, and even nearby highways, pose the greatest health danger.

This level of disparate exposure characterizes African American communities in nineteen states, compared to Hispanic neighborhoods in twelve states and Asian enslaves in seven states. More than 68 percent of African Americans live within thirty miles of a coal-fired power plant[5]—the distance within which the maximum effects of the smokestack plume tend to occur—compared with the 56 percent of whites and 39 percent of Latinos who live in such proximity to a coal-fired power plant.[6]

As longtime Norco, Louisiana, resident Margie Richard told sociology professor Robert D. Bullard: "I am surrounded by 27 petrochemical companies and oil refineries. My house is located only three meters away from the 15-acre Shell chemical plant. We are not treated as citizens with equal rights according to U.S. law and international human rights law."[7]

Houston, We Have a Problem

Bullard, director of the Environmental Justice Resource Center at Clark Atlanta University, has written a dozen books about environmental health in America, including his 1990 classic *Dumping in Dixie: Race, Class and Environmental Quality.* Because of his analyses, we first learned that race, not income, is the single most important factor in the siting of many sources of brain-harming environmental exposures, a revelation which spurred development of the environmental justice movement.

"African American households with incomes between $50,000 and $60,000 live in neighborhoods that are more polluted than neighborhoods in which white households with incomes below $10,000 live," he wrote.

Bullard, then Distinguished Professor at Texas Southern University and founder of the EPA Office of Environmental Equity, explained that the NIMBY ("not in my backyard") sentiment regarding toxic sites that many Americans share has another face: "LULUs." "Race is still the potent factor for predicting where Locally Unwanted Land Uses (LULUs) go. A lot of people say it's class, but race and class are intertwined."[8]

Bullard adds that African Americans' own ability to invoke NIMBY is hampered by their relatively low home-ownership rate. This disparity is driven by racial discrimination, including mortgage redlining. "In 1999," he wrote, "only 46 percent of Blacks in the nation owned their homes, compared with 73 percent of Whites."[9]

Prior to Bullard's work, no rigorous studies by scholars had examined the connection between race and toxic environmental sites. Bullard first demonstrated the primacy of the race connection when he conducted the first U.S. investigation of race and the placement of noxious waste sites. In 1983 the report based on that study, *Solid Waste Sites and the Black Houston Community,* determined that all five of that city's garbage dumps, six of its eight garbage incinerators, and three of its four privately owned landfills were located in African American neighborhoods, although only 25 percent of the city's population was black.[10]

In 1979, Bullard and his wife, attorney Linda McKeever Bullard, sued to stop the siting of yet another municipal landfill in Houston's suburban Northwood Manor neighborhood. Except for the fact that it was over 82 percent African American, this suburban middle-class community was an unlikely location for a garbage dump, illustrating that African American neighborhoods, middle-class and poor alike, were preferential toxic waste sites.

Bean v. Southwestern Waste Management, Inc., which charged race-based environmental discrimination, was the first suit of its kind in the United States, as citizens contested the siting of public facilities in their ethnically distinct communities or neighborhoods. "Without a doubt, this was a form of apartheid where whites were making decisions and black people, brown people and people of color, including American Indians on reservations, had no seat at the table," summarized Bullard.

But the courts decided against the plaintiffs, noting, "An intent to discriminate must be demonstrated. Decisions that may appear poorly based to some people are not necessarily unconstitutional or illegal...." The Supreme Court decided a series of similar cases including *Washington v. Davis* (1976) and *Arlington Heights v. Metropolitan Housing Corp.* (1977), each time maintaining that demonstrating a racially disproportionate impact was not enough: to win, complainants must prove *an intent to discriminate* based on race.

This requirement that the injured communities show intent presented a much higher hurdle than demonstrating the foreseeable disparate impact on ethnic communities. Many saw it as a serious blow to attempts to hold municipal governments and polluters accountable for environmental poisoning of communities of color, but Bullard and other activists he inspired were not deterred.[11]

Why in toxic siting is the disparity so often characterized as being driven by economics or class? This may reflect discomfort with acknowledging U.S. racial harms and race in American culture. But there's also another factor—missing national data. It is easier for many to assuage guilt by entertaining the concept of poisoning hazards stratified by income than by race, because the latter would constitute racism, which evokes feelings of shame in many. Moreover, some definitions of socioeconomic strata include education, which serves to ascribe risk to the undereducated—a subtle form of victim blaming. If studies do not investigate race in the context of environmental hazard placement, it's easier to ignore the phenomenon of race-based siting. If it's not studied, it can't be quantified. As a result, studies tend to focus on environmental hazard placement in the context of class.

Bullard explained to me that "funders simply don't fund studies of race and environmental exposure, so they don't get done." When the epidemiology of race as a driver of environmental exposure rarely enters the canon, race as a risk factor becomes invisible.

Moreover, the focus on socioeconomics in opposition to race is compounded by the mercurial concept of race in popular culture and expression—including within science. Many of us have been taught to think of race as a fixed, genetically mediated, biological characteristic of humans. However, "'race' is chiefly a social category that encompasses what is commonly referred to as ethnicity—common geographic origins, ancestry, family patterns, cultural norms and traditions, and the social history of specific groups," explains Professor David R. Williams of the Harvard T. H. Chan School of Public Health. Medical journal articles and reports that refer to race rarely define it, and when they do, the definitions are inconsistent and vary in their validity. The same is true of government health information sources. For example, Williams points out that racial categories have changed with every census. Furthermore, data on race are collected inconsistently. When race is ignored, economics is focused upon, even if race is the more predictive variable.

Evaluating the nature of the harms done by pollutants is even more complex than determining the relative roles of race and economics. The work of two scientific disciplines is key to analyzing environmental harms to health: toxicology and epidemiology.

Toxicologists use tools like cell cultures and animal models that mimic human disease to test and to tease out the effects of poisons. They seek to control for or eliminate confounding factors like different diets, genetic susceptibility, and even other environmental exposures in order to characterize the nature and strength of a pollutant's health effects.

Epidemiologic studies can identify associations between exposure and harm, but single correlations do not rise to the level of proof. (Multiple well-conducted studies that point to the same culprit, however, can lend power to these correlations.)

We often hear this criticism summarized as "correlation does not establish causation." But what does? The answer varies, but to prove that a pathogen causes a disease, for example, scientists no longer rely on the pat formulae of old, such as the German physician Robert Koch's oft-invoked postulates. His tenets hold that pathogens responsible for a disease must

- be isolated from the diseased host and grown in pure culture,
- cause the disease when a pure culture of the bacteria is inoculated into a susceptible host, and
- be recoverable from the experimentally infected host.

We now know that not every pathogen can be grown in culture, that there are not good animal models for every disease, that not every infected person grows ill, and that even "harmless" bacteria can sicken the immunocompromised or people made susceptible by trauma or surgery that permits bacteria access to their viscera. Today, we must rely upon sophisticated new tools in establishing causality.

Life—And Death—Along the Fence Line

Not all the deadly clouds hover over crowded cities like New York, Los Angeles, and Houston: they also assault suburban and rural fence-line communities that abut toxics-spewing industries, chemical dump sites, and Superfund sites.[12]

According to a 2014 report, *Who's in Danger: A Demographic Analysis of Chemical Disaster Vulnerability Zones,* the percentage of African Americans in the "fence line zones" near chemical plants is 75 percent greater than for the United States as a whole, and the percentage of Latinos is 60 percent greater. These fence-line communities are most often home to people of color, but they rarely receive the media attention of Love Canal, and they are not always poor.

Middle-class African Americans are far more likely than their white peers to live surrounded by belching factories and plumes from dump sites. Airborne exposures are everywhere, but they concentrate in ethnic enclaves. As usual, race rather than poverty dictates the location of Superfund sites, "dirty" industries, and their ilk.

Triggered by their proximity to polluting industry and dump sites, African Americans and Hispanics, the largest minority groups in the United States, suffer triple the U.S. asthma death rate for whites. We know that dust mites, pets, tobacco smoke, cockroaches, and mold[13] are among the risk factors for asthma, but poor external air quality drives not only asthma but also cancer rates, which are higher among African American and Hispanic populations.

Noxious airborne toxins inflame blood vessels, including

those in the lungs, which produces life-threatening respiratory disease. Rampant brain damage strikes adults and children alike thanks to airborne heavy metals like lead and mercury as well as hydrocarbons in fuel exhausts and industrial emissions. These directly threaten intelligence by impairing neurological function. And as if all this were not dire enough, I was surprised to learn from global studies that the asthma driven by pollution itself causes a loss of intelligence and a reduction in IQ.

Losing Breath and Brain: The Scourge of Air Pollution

In relatively affluent and white areas, policing of the environment is visibly stringent. In wealthy, white neighborhoods of New York City, trash collection is frequent and complete, and even noise and pet pollution is diminished by signs citing high fines for horn blowing, pet droppings, and dumping. These are warnings that do not appear in areas of the city populated by the poor and people of color.

Disproportionate risks for communities of color hang in the very air. Our vulnerable brains are awash in chemical threats, but national data tell us that a largely invisible, intangible culprit tops the list of hazards: air pollution. Its toxic components damage even our intelligence, lowering our IQs.

These communities are far from alone. The World Health Organization (WHO) found that more than four of every five urbanites on the planet, most in the developing world, live in neighborhoods where air quality falls below minimal health standards.

Cities like Karachi, Lagos, and Beijing[14] are notorious for their visible smog, which shrouds their citizens in a witches' brew of poisonous chemicals and brain-draining particles.

But the relatively clean air of the United States also features plumes of pollution that impair health. According to *Nature,* air pollution kills 55,000 Americans annually.[15]

An October 2017 report[16] in *The Lancet* identified air pollution as the number one cause of pollution-related illness and death worldwide. Gases like carbon monoxide (CO), sulfur dioxide (SO_2), nitrogen dioxide (NO_2), and ozone (O_3) are one component of air pollution. "Particulate matter"—vanishingly tiny suspended solids that threaten human well-being—is another.

The developing brains of children are the most dramatically injured because they have a greater lung surface area relative to their body size, giving them a greater relative exposure to noxious gases and suspended particles than adults. Fetuses and infants fare worst of all.

African American asthma rates are driven in large part by living, working, and studying in toxics-laden environments. The greatest proportion of pollution-exposed African Americans live within half a mile of the active oil wells, gas wells, and processing plants of Texas, Ohio, and California. The next highest proportion lives in Louisiana, Pennsylvania, and Oklahoma, where industries violate the often-inadequate EPA standards for air quality.[17]

Doris Browne, M.D., president of the National Medical Association, told NBC News that the effects of this pollution include 138,000 asthma attacks annually in school-age children. "It's a significant problem and we should all be concerned by these health disparities."[18] The report added that black communities in Chicago, Washington, D.C., and New York City were also targets because airborne pollutants disperse for miles before becoming ozone smog.[19]

Asthma may grow milder with age, but it doesn't always: This lifelong malady kills adults, like actor Moses Gunn and newscaster Harold Dow, as well as children. African Americans account for 13 percent of the U.S. population, but 26 percent of asthma deaths.

Although animal dander and dust mites are known to trigger asthmatic attacks, studies revealed that living in homes infested by cockroaches also elevates risk. Children also encounter these vermin in antiquated schools where racial minorities are more likely to spend their days because school racial segregation is worsening rather than abating.

IQ and Oxygen

But what has the heightened African American asthma rate to do with lowered intelligence and depressed IQ scores?

Everything. People with asthma suffer episodes where they struggle to breathe, sometimes for very long periods. In so doing they often experience hypoxia, the deprivation of oxygen to the brain. If this continues for too long, asthmatics, near-drowning victims, and others who suffer hypoxia can experience lifelong aftereffects, including "lower neuropsychological performance," according to Harvard researchers who studied perinatal exposures and later cognition.[20] They write that "a significant impact on multiple behavioral and cognitive outcomes" was found in newborns who had suffered hypoxia when they were tested at age seven. This included a decreased verbal IQ.[21]

Air pollution doesn't lower intelligence only through triggering asthma. The University of California's Anthony S. Wexler and Pamela J. Lein write that other aspects of air pollution cause a legion of brain disorders, including "degenerative disease, in particular, Alzheimer's disease (AD) and diverse neurodevelopmental disorders, including autism spectrum disorder (ASD), attention deficit hyperactivity disorder (ADHD), learning and intellectual disabilities and schizophrenia."[22]

In recent decades, new tools and closer scrutiny have allowed us to see how diesel fuel residues and air pollution directly damage the brain and lower intellect.[23] A 2008 Harvard School of

Public Health study[24] of 1,000 pregnant Boston women who carried backpacks to measure their air-pollution exposure until they delivered found that they were constantly exposed to about 0.53 μg/m3 of black carbon, exposures associated with intellectual decline.[25] When these women's children underwent a battery of cognitive tests eight to eleven years later, these exams revealed measurable decreases in verbal and nonverbal intelligence as well as in memory. They found lowered scores consistently, at all levels of exposure. These linear regression–based analyses* did not establish causation, but this is strong evidence.[26]

A subsequent study found that children around age ten who had been exposed to air with high levels of black carbon (soot) "suffered a decrease in cognitive function across assessments of verbal and nonverbal intelligence and memory constructs."[27] The scientists concluded that particulate matter—tiny airborne pieces of various carbon compounds and heavy metals as a result of burning fuel—was largely responsible.

Referencing both studies, the Harvard scientists confirmed that "ultrafine particles can reach the brain... [raising] the question of whether traffic particles can have neurotoxic effects."

The resulting degree of cognitive and memory impairment is comparable to that caused by other environmental neurologic poisonings. A 10 microgram per deciliter (μg/dL) increase in blood lead is associated with a loss of about five IQ points. Children whose mothers smoke moderately have an average decrease of four IQ points. In the group of Boston children, a 0.4 μg/meter^3 (0.4 micrograms per cubic meter) increase in airborne black carbon predicted a three-point decrease in IQ.[28]

Few of us who are nonscientists can meaningfully compare

---

* Regression is used to find out what equation satisfies a group of data points. Linear regression measures the relationship between a scalar variable called *y*, for example, IQ, and an independent variable, for example, air pollution, denoted *x*. It can be predictive when deployed in a valid model.

these exposures—how does 10 µg/dL of blood lead compare to 0.4 µg/m^3 of inhaled carbon, for example?

But we *can* see that for these common exposures, the drop in IQ range is similar: around three to five points. As laid out in Chapter 2, a five-point drop in IQ is *not* trivial. It has serious individual and societal implications, including the ability to drag down the national average IQ, the number of intellectually gifted persons, and the income of the entire nation.

Elevated carbon dioxide (CO_2) levels also lower intelligence when they impair breathing, inducing oxygen deprivation and often triggering asthma.

So befouled air degrades cognitive development and brain function, depressing the IQ three to four points in some studies. But studies also show that such oxygen deprivation also induces anxiety, depression, and suicide as well as lowered intelligence.

Magnetic Malady

Scientists don't understand every specific route of air-pollution injury. But *Environmental Health Perspectives* reports one known mechanism by which "particulate matter" like black carbon from the incomplete burning of fossil fuels injures the brain.

It consists of millions of tiny spheres of several carbon forms including, for example, magnetite and iron oxide, better known as rust. These are already known to cause preterm births and disability. Although they are imperceptibly small—one must join 250 of these nanospheres to achieve the thickness of a human hair—"these particles are made out of iron, and iron is very reactive, so it's almost certainly going to do some damage to the brain," explains Professor David Allsop, an Alzheimer's specialist at Lancaster University.[29]

When it comes to Alzheimer's, air pollution has become a prime suspect. Recently, laboratory studies have suggested that iron particles like magnetite contribute to the disease's characteristic protein plaques. In the journal *F1000Research,* Soong Ho Kim reports that the characteristic amyloid plaques of Alzheimer's quickly appeared in mice after they were exposed to tiny components of polluted air, called nickel nanoparticles.[30]

When people who died in heavily polluted Mexico City were autopsied in 2004, amyloid plaques and inflammation were found throughout their brains. The tiny particles of magnetite in air pollution have also been linked to dementia and to Alzheimer's by other U.S. studies.[31]

As *Time* magazine notes, the plaque-and-inflammation-affected populations tend to be poor, black, and Hispanic ones concentrated in low-income areas.[32] African American rates of Alzheimer's are as much as 100 percent higher than those of whites, constituting what the Alzheimer's Association calls a "silent epidemic" among black Americans. This type of pollution may help explain why, and understanding its role in causing the disease may present a route to a cure. However, scientists writing in *Current Alzheimer's Research* think that a potential solution exists now: prenatal choline supplementation, which is also touted as a potential treatment for Down syndrome and as a preventative measure against fetal alcohol spectrum disorder.[33]

Airborne nanoparticles of magnetite cause all this neurological mayhem because they literally invade the brain. This ore, more magnetic than any other natural mineral in the world, is unleashed into the air by diesel-burning vehicles.[34] When polluted air is inhaled, magnetite travels from the nose to the brain along the olfactory nerve. Animal studies also indicate that particles can migrate from the upper respiratory tract to the nervous system and the brain.[35]

Skeptics who question magnetite's danger to cognition point out that it occurs naturally in the human brain and therefore is unlikely to be an agent of harm. Magnetite *is* found naturally in the brain, but "natural" and "safe" are not synonyms.[36]

In 2016, scientists found an abundance of human-made magnetite in about forty representative samples of human brains from Mexico City and Manchester, England. They knew the magnetite came from air pollution because naturally occurring magnetite and the pollution-borne type—the one suspected of causing disease—are quite different. Intrinsic magnetite is jagged and crystalline, but the high heat of industrial engines produces "pollutant" magnetite that is smoothly rounded, and this foreign type is present in the brains of pollution-affected people in far greater quantities—as much as one hundred times the amount of the naturally occurring form.[37]

A research study described in *Scientific American* also links Alzheimer's to DDT, and suggested that genetic vulnerability may combine with DDT exposure to create the most devastating cases.[38]

Air pollution damages the brains and undermines mental abilities of adults, too,[39] and even the elderly. Cognitive damage was measured in older women from both rural and urban environments with long-term exposure to air pollution from heavy traffic, and it was found to be cumulative, increasing over time. Of course, in order to interpret this damage, researchers had to correct for many confounding factors. These included sporting activities, age, education level, depression, chronic obstructive pulmonary disease (COPD), chronic vascular disease, heart disease, stroke, high blood pressure, and diabetes.[40] Having done so, they were confident in concluding that air pollution poses a threat to the intelligence of the elderly.

Petrochemical Corridors

According to American companies' own Toxic Release Inventory filings for 2010, 21,000 U.S. facilities reported discharging 3.9 billion pounds of toxic chemicals into U.S. land and air.[41] In

Louisiana alone, 361 industries released 130 million tons of hazardous wastes and emissions, fully 64 percent of which was dumped into the black parishes that are home to cities like Alsen, which is 98.9 percent African American. The industry calls this "the petrochemical corridor," but for its residents, the plethora of excess disease has earned it the sobriquet "Cancer Alley." Bullard adds, "The playgrounds in Norco, La., which sits in Cancer Alley, are across from a huge Shell refinery. You stand there 15 minutes, and you can't breathe."

Moreover, industries have dumped 21.8 billion pounds of industrial waste into the water,[42] and African American, Hispanic, and Native American communities are in closest proximity to these toxics-spewing industries.[43] Native American communities in particular, which often lack access to potable water, basic health care, or even electricity, are plagued by waterborne pollutants, poisoned fish, and coal-fired energy plants that disgorge mercury. One of every ten U.S. power plants sits on Native American land.

In this chapter, I describe a few of these poisoned communities and how they have been plagued by intellect-sapping toxic exposures. But first, I recount a mid-1980s encounter that brought home to me the realization that some U.S. industries poison unsuspecting Americans with impunity.

Manufacturing Confusion: Benzene, Sulfides, and Fence-Line Communities

I met Cecil Fisher* only once, but I will never forget him. His brother Eric was a fellow movie buff who, as we caught up between show times, told me that Cecil had nearly died of leukemia.

---

* The names of Cecil and Eric, and other details of this episode, have been changed to protect their privacy.

As Eric and I stood beneath a theater's marquee in Rochester, New York, early one summer evening, a man approached us with a strange gait, unsteady yet rapid. His skin was coarse, his face gaunt, his eyes sunken, and his pale cotton shirt billowed around his precariously thin form. His crinkly reddish-brown hair was so sparse that his entire scalp was visible. Smiling, he drew closer and clasped hands with a beaming Eric, who exclaimed, "Cecil—you're walking without the cane: All right!" Only then did I realize that this was not a wizened older man but Eric's twentysomething brother.

As we chatted about our lives and jobs, Cecil mentioned that he had felt well enough to begin job hunting but decided to postpone it on his doctor's advice. "I feel okay, but I have trouble filling out the forms and remembering the simplest things. I felt like an idiot when I couldn't remember my birthdate or the name of my high school, and they wouldn't let me take the form home with me. My doctor says memory problems sometimes happen and I should give it time." He paused. "I'll need a new job, though."

Cecil had recently been dismissed from the furniture factory where he'd worked for years, but said he was glad because his job had been to stand inside a vat containing benzene, dipping chairs into it, and he hated the smell. "It gets into everything."

I was appalled. "Benzene—B-e-n-z-e-n-e?" I asked, hoping that he meant *benzine,* a word with which it is easily confused. "Are you sure?"

"That's what the sign says."

The brothers continued to laugh and banter, but I was too stunned to hear another word. I worked in a poison control center, and I knew that benzene can cause leukemia.

I moved away soon after and lost touch with his brother, so I don't know whether Cecil recovered. But I now know that surviving cancer may not have ended his medical problems,[44] and his memory problems may have stemmed from his benzene exposure as well. According to the CDC, benzene causes more than cancer, blood diseases, and impaired reproduction.[45] It also assails the brain with neurological and cognitive effects like short-term drowsiness, convulsions, confusion, and mental impairment. These thinking problems may not abate with time.[46]

In fact, a *Toxicology* report based on the study of 2,143 utility workers found that "high exposure to solvents was significantly associated with poor cognition; for example, those highly exposed to chlorinated solvents were at risk of impairment on the Mini-Mental State Examination." Moreover, pregnant women face a dual risk: one to their own health, and one to the health of their fetuses.[47] Benzene can also induce neurobehavioral changes in babies that lead to cognitive damage.

Benzene plagues workers beyond the urban factory where Cecil was employed. In 2015, Tonawanda Coke, of Tonawanda, New York, was ordered to pay $12 million in civil penalties for violations of the Clean Air Act after its failure to follow safety regulations resulted in releases of coke oven gas, which contains benzene and other harmful chemicals.[48]

In all, more than 6.7 million African Americans—who constitute 14 percent of the national population—face toxic exposures to benzene and sulfur dioxide emissions from oil refineries in ninety-one counties.[49] "Fumes Across the Fence-Line: The Health Impacts of Air Pollution from Oil and Gas Facilities on African American Communities," a 2017 report by the Clean Air Task Force (CATF) and the NAACP, found that more than 1 million African Americans live within half a mile of an oil and gas operation.[50]

This is not news to pioneering environmental sociologist Robert Bullard, father of the environmental justice movement. Since the 1990 publication of *Dumping in Dixie,* his first book on environmental racism, he has decried the high disease rates plaguing fence-line communities of African Americans.

Jobs or Health?

Some areas, like Houston, have no zoning laws to protect residents from sharing their space and air with toxic industrial effluents. Other municipalities, especially but hardly limited to poor Southern towns and cities, accept these undesirable industries because they want the jobs and the tax base they bring with them. Politicians and industries often seek to justify the unpleasant and dangerous presence of polluting companies by arguing that they provide jobs that give otherwise unemployed poor residents of color better lives. A dirty, even hazardous job in a polluting industry, they imply, is better than no job at all.

But studies such as a 2011 report in *Occupational and Environmental Medicine* have dismantled this argument. In addition to the direct exposures that can cost workers and their families their health and lives, the *OEM* study of 7,000 workers documented that such jobs destroy mental health as well. People in poor-quality jobs where high demands are coupled to low autonomy and rewards are out of synch with effort suffer far worse mental health from the malignant stress than do the unemployed.[51] Moreover, polls show that these communities are unwilling to trade jobs for unhealthy environments.[52] "African Americans support clean energy, clean jobs, and clean power plants, with 83 percent support in favor of setting limits on carbon pollution from coal- and gas-fired power plants in concert with the Clean Power Plan's standards that the Environmental Protection Agency finalized in August," concluded a 2015 CleanTechnica poll.[53]

Nonetheless, Bullard's publications continue to document how these disparate racial exposures have grown rather than abated, and Marcus Franklin, author of the "Fumes Across the

Fence Line" report, agrees. "There's a growing threat faced by 'fence-line' communities."[54]

Anniston Apocalypse: PCBs Are Unleashed on a Company Town

Shirley Baker, a nurse, deftly ties on a surgical mask before opening the door. But she's not striding into the operating room: she is about to mow her lawn, which is ringed by a high chain-link fence festooned with biohazard signs.

Baker lives in Anniston, Alabama, sixty-three miles from Birmingham. Her city of 24,000 is 52 percent African American, but mostly it's the city's black residents who inhabit the neighborhoods that have fallen into decay wrought by widespread pollution. These form a patchwork quilt of moribund communities and biological "dead zones" where nothing grows.

Behind some doors, the unemployed fight cancer, paralysis, memory loss, and a bewildering array of poorly characterized diseases. Subdued children play, eerily quiet, against a backdrop of toxic lawns, oily creeks, tainted vegetation, and sere trees.

Other neighborhoods are already dead. Vegetation has overtaken blocks of abandoned houses, with streetlights gone permanently dark, empty churches, and, always, the biohazard signs. Everywhere in Anniston, worried parents shoo children from parks and playgrounds. Many backyard creeks run blood red. Homeowners have forsaken their poisoned gardens to grow greens in sterile plastic buckets.

Children here seem slower and sicker than most, says Shirley's husband, David, whose own brother died at sixteen after years of illness. His daughter displays an assortment of behavior problems and has been relegated to special-education classes. A ten-year-old girl down the street has uterine cancer, he says, and he repeatedly assured me that several children nearby have been born with "two brains." I wonder whether he means that they were born without a normally developed corpus callosum, which separates the brain's hemispheres, but I can't find out: their parents don't return my phone calls.

For many Americans, a modern dread of contamination has been distilled into cathartic postapocalyptic film fare such as *28 Days Later, Right at Your Door,* or *Dawn of the Dead* that feature poisoned lands and communities thronged with those so degraded by infection and environmental exposures that they have lost their intelligence and even their humanity. Confronting these cinematic horrors allows us to share a benign frisson of fear, secure in the knowledge that when the lights come up, we'll emerge into normalcy.

But for Anniston, the apocalypse is all too real—and for most, inescapable.

Once, from 1929 to 1971, Anniston was a company town. First, the Swann Chemical Company produced polychlorinated biphenyls (PCBs) there. Then Monsanto Industrial Chemicals took over the plant in 1935.[55]

The $20 billion[56] corporation Monsanto, which brought the world the sweeteners saccharin and aspartame, boasts versatile chemical production, a checkered past, and a highly controversial present. Monsanto has produced synthetic fibers, plastics including polystyrenes, pesticides, and agrichemical products. It has also acquired many chemical and electronics companies that make products as varied as aspirin and light-emitting diodes (LEDs). Just last year, it merged with the equally rich and powerful Bayer Corporation.

Monsanto first devised or marketed DDT, dioxin, 2,3,7,8-tetra-chlorodibenzodioxin (TCDD), 2,4-D, 2,4,5-T* PCBs, other

* Colorless, crystalline, tasteless, and almost odorless organochlorine compounds used as insecticides whose health and environmental impact was indicted by Rachel Carson's book *Silent Spring* in 1962.

halogenated hydrocarbons that are carcinogenic even in small doses, and dioxins.

In 1960 an "agricultural" division was established, which trafficked in the hazardous defoliant Agent Orange[57] as well as the controversial recombinant bovine somatrophin* hormones used to increase the milk yield of cows.

By 1969, the plant was Anniston's major employer, discharging 250 pounds of PCBs into Snow Creek, at the heart of the city's black residential community, every day.[58] PCBs are "brain thieves" that erode the structures and functioning of the brain and nervous system, and they are also *endocrine disruptors* that impede the healthy physical and mental development that is normally guided by hormones (see Chapter 4). Although the company and its apologists insisted that one can tolerate significant amounts without ill effects, this reassurance rang hollow in Anniston neighborhoods that found themselves suddenly battling a legion of ailments from cancers to memory loss, confusion, and a slew of other intellectual problems. Children's behavioral problems snowballed in Anniston, along with rates of attention-deficit disorder and poor school performance.

The news media often focus their outrage on cancer clusters and visibly crippling lung and mobility ailments caused by PCBs. But PCBs' most persistent legacy is the invisible harm they wreak on the brains of the young. Industry and some media accounts downplay small exposures as innocuous—claiming that "50 bathtubs' full" of PCBs at low concentration is required to do harm—but this has not been proven.[59] In reality, extremely small amounts of PCBs harm the developing nervous systems of fetuses and children. Even very low concentrations are harmful for immature brains during their critical windows of development. In 2000, researchers calculated that a PCB concentration

---

* BSTs are banned in European Union countries as well as Canada, Australia, New Zealand, Japan, and Israel.

of just 5 parts per billion in a pregnant mother's blood can have adverse effects on a developing fetal brain, giving rise to attention and IQ deficits that appear permanent.[60] Five parts per billion is equivalent to one drop in 118 bathtubs full of water.

Anniston Apocalypse

David Baker, Shirley's husband, is one of many who have borne witness to the painful toll of PCB poisoning. Growing up, he and his brother Terry would play in the neighboring woods and rivers, exploring, shooting arrows, and splashing in creeks that ran with water containing PCBs. But their bond was severed in 1970 when Terry sickened dramatically and died of a spectrum of diseases

Anniston, Alabama, environmental activist David Baker stands over the grave of his brother Terry, who died at just sixteen, felled by lung cancer, a hardening of the arteries, and a brain tumor—an assortment of environmentally triggered illnesses. (Courtesy of Mathieu Asselin)

that are usually associated with aging: lung cancer, hardening of the arteries, and a brain tumor. He was sixteen years old.

Denise Chandler, forty-six, has also had a front-row seat to death caused by reckless pollution. She, too, regularly played with her brother in one of the neighborhood's chemical-imbued ditches. "We floated our little boats in it and waded in it, but we didn't know it was loaded with PCBs," she recalled. Decades later, when they were finally tested, they discovered that they both had high blood PCB levels. She suffers from sarcoidosis, an autoimmune disease characterized by widespread tissue inflammation. Her brother suffered myriad health problems before he died of kidney failure at age forty.

But their problems are more than physical: two of Chandler's three children were diagnosed with learning disorders.

From 1935 until 1971, without warning its neighbors, Monsanto disposed of tens of thousands of pounds of PCBs by dumping them into creeks or burying them in and around Anniston. But industry wasn't the only source of Anniston's chemical exposures: in 1917, the military established Fort McClellan there, and the U.S. Army manufactured and trained heavily with them from World War I until its closure in 1999.

Despite being saturated with environmental poisons, Anniston became an icon of normalcy when it was named the "All American City" in 1978. The very next year, Snow Creek began to run red, heightening residents' suspicions about the effects of chemical exposures in their communities. In 1996, one of the dumps started leaking. It was then that residents began learning the extent of their contamination. Years of unchecked chemical dumping had utterly poisoned the lands of Anniston's black neighborhoods.

A former union organizer, David Baker took action, channeling the pain of his brother's loss into ensuring that the citizens of Anniston receive justice. He created Community Against Pollution in 1998 to force the chemical companies to clean up the contamination and compensate those harmed by it. At his urging,

This depot stored an array of toxic industrial chemicals in Anniston, Alabama. (AP Photo/Dave Martin)

the EPA tested Anniston's soil and water as well as the blood of its residents. It was alarmed to find that the blood of Anniston's townspeople had the highest recorded levels of PCBs *in the nation.*

But their complaints drew a desultory response from industry, so Baker organized residents, who filed a number of lawsuits against the unresponsive Army and against Monsanto.

As the evidence of Anniston's poisoning mounted, Monsanto shed its industrial-chemical fibers business into a separate company called Solutia. It also began trying to buy up heavily tainted properties, including a local church. This further fueled residents' suspicions that despite its denials, the company had long known how dangerous its dumped chemicals were.

They were right. In 1966, Monsanto had hired the late Mississippi State University professor Denzel Ferguson to investigate the health effects of its PCB pollution in Anniston. When Ferguson's team of biologists lowered bluegill fish into the city's creeks to monitor the water's health effects, all twenty-five fish died within three and a half minutes. "It was like dunking the fish in battery acid," one team member told the *Washington Post.* "I've never seen anything like it in my life," said another. Yet Monsanto ignored the biologists' urgings to warn residents and clean up the waters.

Monsanto did respond with alacrity to concerns about the fiendish toxicity of PCBs voiced by Swedish scientists the very next year. A Monsanto official wrote Emmett Kelly, the company's medical director, beseeching him, "Please let me know if there is anything I can do so that we may make sure our business is not affected by this evil publicity." In 1979, Monsanto closed the Anniston chemical factory.

Wake-up Call

In 2002, the people of Anniston suddenly learned from a *60 Minutes* investigation that theirs was one of the most toxic cities in the nation. PCBs are widely disseminated in industry products, so widely in fact, that the average American has PCB blood levels of 2 parts per billion (ppb).

But the mostly black victims of Anniston suffered huge exposures. Howard Frumkin, M.D., told me, "Anniston has the highest levels of PCB exposure of any town in America, of any town that I've ever heard of." David Baker has PCB blood levels of 341 ppb. Most of the residents know their levels.

As in most toxic communities, the PCBs in Anniston were not acting alone. Anniston is blanketed by a mixture of asbestos, arsenic, and other unstudied chemicals. Even if their effects had been known, mixtures of chemical exposures can act in an unexpected manner. One chemical might mute or potentiate the effects of others in an additive, or even a synergistic, manner.

Lead, cadmium, arsenic, and mercury are pollutants that are commonly found together, and all have long-lasting effects on the brain. According to a 2016 study published in *Environmental Toxicology Pharmacology,* the metals share many common pathways for causing cognitive dysfunction, and all bind to a particular receptor, which makes their effects synergistic: harmful exposure to the mixture results in a greater poisoning than the added effect of poisoning by each of its members.[61]

Anniston residents filed class-action suits against Monsanto Chemical, alleging that it had knowingly dumped PCBs into the local water supply for decades. Like the Anniston Army Depot, Redstone Arsenal, and other industries, Monsanto had also exposed residents to asbestos, which can cause diseases like the deadly lung cancer mesothelioma.

Residents settled a case against the company in April 2001 for $43 million, and in 2003, a jury determined that the Anniston Monsanto plant had imbued Anniston with PCBs. Monsanto and Solutia agreed to pay $600 million to settle the claims, but Solutia declared bankruptcy that very year.

Its forty-year monopoly on PCBs left Monsanto with a long history of injury and abandonment that was revealed to the world when incriminating Monsanto documents were posted on the *Chemical Industry Archive* website as a result of the lawsuits.[62] The documents indicate that Monsanto long knew of the severe damage it caused by dumping millions of pounds of PCBs into Anniston for four decades.[63]

In 2003, the Department of Defense began destroying the chemical-weapon stockpile, including nerve gas, stored at the Anniston Army Depot.

Most of the settlement funds went to lawyers and cleanup efforts, leaving the people of Anniston unemployed, impoverished, sick, mentally hobbled, and, in many cases, dying. They were unable to sell their homes and flee. The Army and EPA had failed to protect them and they were unsure where to turn next.

David Baker, now the executive director of Community

Against Pollution, was sure of one thing: they needed a powerful champion. He found one: Johnnie Cochran.

Baker approached Cochran, who was fresh from securing O. J. Simpson's not-guilty verdict, and after hearing stories of Anniston's poisoning and rampant illness, Cochran agreed to help them obtain compensation for their property losses and a health clinic to address their cancer, liver disease, and mental problems.

As he denounced the failure of the government to protect Anniston residents, Cochran declared, as I noted in this book's Introduction, "There is always some study, and they'll study it to death, then thirty years later, you find out it's bad for you.... We know it's bad for us right now!"

Cochran's class-action suit procured the largest settlement ever won in the United States. The victims of Anniston won $300 million from Monsanto, its subsidiary Solutia, Pfizer, and other firms, none of whom admitted any wrongdoing. Fifty million dollars was reserved for a health clinic to address the medical aftermath of Anniston's poisoning.

Anniston's Aftermath

The celebration was short-lived. About 47 percent of the settlement—some $142 million—went to the 18,447 plaintiffs. Adults received an average of $9,000 each, and each child $2,000, an absurdly low amount considering that each faces a lifetime of disability, including reductions in IQ, that erode earning ability. With the exception of the portion of the settlement set aside to fund the health clinic, the rest of the money went to pay the lawyers.

The townspeople's expectation of life-altering financial compensation evaporated, and most remained trapped in their tainted homes. The health clinic ran out of funds and closed in 2017.

In comparison, victims of Japan's Minamata poisoning, caused by the release of methylmercury from the Chisso Corporation's chemical factory from 1932 to 1968, reaped $1 billion in compensation, an average of $20,000 per person.

Today, Anniston's community meetings are thronged with the sick, who tend to introduce themselves by ticking off their sky-high PCB levels and bestiary of diseases. Even the relatively young compare their cancers and frailties. Baker speaks of filing another lawsuit, but without his old fervor; he has received anonymous death threats from neighbors who felt betrayed by their low compensation from the earlier settlement. Even an outsider can see that their solidarity is diminished.

"Monsanto did a job on this city," summarizes Opal Scruggs, sixty-five, who, like everyone else in her neighborhood, has elevated blood levels of PCBs. "They thought we were stupid and illiterate people, so nobody would notice what happens to us."

Arsenic and Old Waste

Around 1962, the city of Fort Myers, Florida, began searching for land to buy in its historically black Dunbar community, whose racial demographic was long maintained by de jure, and later, de facto, segregation. Residents and other property owners were asked to sell their lots "for municipal purposes": the city said it would use the purchased property to build affordable housing in a development called "Home-a-rama." Although nothing was ever built, the name stuck.

Instead, the city dumped waste from its water treatment plant in the site's ponds and grounds, beginning with 25,000 cubic yards of sludge in pits extending deeper than the water table. The site bore no identifying signs, was unfenced, and was surrounded by African American families whose numbers soon grew explosively in a building boom.

"All this time, we bathed in that water, we cooked in that water, kids played in that water. We played cops and robbers in there, hide and seek, built club houses and played in the trees," said Shanon Reid, one of a family of fourteen who lived in Dunbar during the 1970s. "In certain spots it was like soft clay. We called it 'orange slide' because it was squishy and slimy."[64]

Curtis Sheard also grew up near the dump in Dunbar, playing with his friends in "the orange slide."

Milton "Shorty" Johnson recalls that land across the street from his home had what they called "quicksand"—lime sludge that had been dumped there by Fort Myers in the 1960s. A 2006 EPA official's photo of a child's toy atop the dirt beneath which arsenic-laden sludge festered attests to the fact that children still played at the contaminated Home-a-rama site—which still bore no fencing or warning signs. Only in 2017 did the city finally erect a handful of no-trespassing signs.

But the residents did not learn until 2007 that arsenic and other toxic substances had been dumped regularly in Home-a-rama, the heart of their community, for nearly five decades. Ten micrograms per liter is the acceptable standard for arsenic in drinking water, but levels on some parts of the property have ranged from 11 to 22 micrograms per liter—more than twice the limit. In 2012, tests showed that the toxic sludge dump exceeded the EPA safety levels by as much as *five times.*

Arsenic is a human carcinogen, and people are exposed via many industrial routes, like coal-burning factories and mines. Some people are exposed by ingesting soil or by eating unwashed produce. People affected with pica, a condition that makes one crave nonfood items like dirt or chalk, are also susceptible. Arsenic is even a component of tobacco smoke. But exposure most often comes from tainted water, as it did in Dunbar.[65]

Sheard, who ran for mayor in 2015, speaks for many when he decries the lack of transparency. "We have a water-line break and the city is responsible to notify the public in forty-eight hours, but no one notified the Dunbar community of the toxic sludge."

"That's something I wasn't aware of happening in my own backyard," Dunbar's Jumar Hillards agreed in an interview for WINK News.[66]

"Is this happening in every neighborhood? No. It just seems like the low- to moderate-income black, brown people constantly get the short end of the stick," Dunbar resident Crystal Johnson said.

"Whatever the garbage and junk, whatever they don't want, they give to Dunbar," Johnson said.

"Those people, they didn't care what happened out here [in Dunbar]," says retired truck hauler Clarence "Pappy" Mitchell, who worked for the city of Fort Myers in the early 1960s and often dumped toxic sludge in the Dunbar section. "This was a dumping area."[67]

Mitchell recalls that when a resident asked for coquina shells to harden the neighborhood's dirt roads, Mitchell's supervisor responded, "Hell, ain't nobody live out there but a bunch of niggers. Take a load of sand instead."

If you're tempted to believe that such sentiments died with racial segregation, reflect that segregation has never ended. In fact, it has worsened since the time Mitchell recalls, thanks to the widening adoption of conservative policies.

According to Professor David R. Williams, if there has been any decline in segregation, this has had no impact on the very high percentage of "black" census tracts. This residential isolation of most African Americans, along with the fact that the concentration of urban poverty remains so high, means that racial parity could be achieved only if 66 percent of the black population moved to nonblack areas: economists call this extreme level of segregation *hypersegregation*.[68]

A study of the 171 largest cities in the United States concluded that there is not even one city where whites live under equal conditions with blacks. "And the worst urban context in which whites reside," avers Williams, "is better than the average living conditions of blacks.... One of America's best-kept secrets is how residential segregation is the secret source that creates inequality in the United States."

Its actions suggest that the city of Fort Myers knew how dangerous the sludge was. "In 1994, they tried to sell [the polluted land] to Habitat for Humanity. In 2001, they tried to say it had no sludge. The City of Fort Myers' attempts to pass off the toxic Home-a-rama site in Dunbar go back farther than officials have publicly said," wrote Patricia Borns in the *News Press*.[69] In 2002, the city asked for a site assessment to provide liability protection against any hazardous waste cleanup. Fort Myers then declared that "no sludge" existed at the site, but the assessor, Steve Hilfiker, insists he had not given the site a clean bill of health.[70]

In a videotaped interview, Hilfiker qualifies that his was only a "preliminary assessment." He went on to say, "we did not identify anything on the ground surface but... subsurface testing would be necessary," and cleanup would indeed be needed to "excavate and properly dispose of material."[71]

The children who once played in the pond every day are now middle-aged and worried. They demand answers. When I spoke with Tangela Rodgers, she asked, "This coming out now, after forty years? How many lives has it affected? Why dump it here, in our neighborhood?"

Testing in 2008 showed unacceptably high arsenic levels, but regular groundwater testing was delayed for two years because of municipal foot-dragging, according to the Department of Environmental Protection (DEP). "Approximately 700 days elapsed without a written response to our inquiry by the responsible party (the city)." In 2017, Fort Myers renewed its denials when the mayor claimed, "I see no reason for residents to be concerned. There's no evidence that they should be concerned."[72] But that year, the city finally erected a handful of no-trespassing signs at the site. Dunbar native Almeda Jones says she still sees children playing in the vacant lot and that the city needs to clean it up. "It's not good for your health."

She's right. As the National Institute of Environmental Health Sciences notes, arsenic in drinking water is a significant and well-established environmental cause of cancer.

But less well known is the catalog of arsenic's neurological effects, which have been unearthed by scientists around the world. Arsenic is able to invade the developing brain where its effects are toxic. Not only does the blood-brain barrier (BBB), whose function is to shield the brain from harmful exposures, fail to keep it out, arsenic itself can weaken and disrupt it.

In animal models, perinatal arsenic exposure shrinks the brain, reducing the numbers of neurons as it distorts the activity of neurotransmitters.

Fifteen epidemiological studies of humans indicate that early exposure is associated with lowered intelligence and reduced memory.[73] Even worse, exposures below the allowable limits trigger these brain effects. Some neurocognitive consequences become apparent only later in life, and exposures to other chemicals and the timing of the exposure all seem to affect the intellectual consequences.

As is often the case, infants and children seem more vulnerable, probably because of their relatively greater consumption of food and water. However, arsenic is not excreted in breast milk, so breastfeeding may offer some protection by replacing the consumption of tainted water and food.

The federal Agency for Toxic Substances and Disease Registry determined that acute toxic exposures to inorganic arsenic have been shown to lead to emotional lability and memory loss.[74]

Low-level and chronic exposure to arsenic is also associated with serious effects on intellectual function across a broad age range, according to about twenty studies conducted around the globe. These investigations used a wide range of cognitive tests—about fifty—including variants of the Raven Colored Progressive Matrices Test, Wechsler Intelligence Scale, and the DSM-IV.[75]

Seventeen epidemiological studies assessed for neurocognitive or behavioral outcomes, and of these, fifteen showed neurocognitive or intellectual deficits associated with arsenic exposure, while two failed to show these effects.

A meta-analysis of arsenic-exposed children also indicated intelligence deficits; the overall mean IQ score of children who lived in arsenic-exposed areas was more than six points lower than that of unexposed children.

Adolescents who had been exposed to arsenic-contaminated water early in life performed more poorly in three of four neurobehavioral tests compared with unexposed controls.

Even the elderly suffer from long-term exposure to arsenic. A study of a geriatric population showed that long-term low-level arsenic exposure, even at levels below the current safety guidelines (10 μg/L in adults), was significantly associated with "poorer global cognition, diminished visuospatial skills, reduced language, slower processing speed, impaired executive functioning, and diminished short-term memory."

One of these elderly people is Almeda Jones's sickly brother, whom she cares for. She worries that his health woes may be tied to arsenic exposure from the "smelly" water in his backyard that has been "gushing, coming out of the ground... They [city of Fort Myers] must take care of it."

Why don't Dunbar residents simply move? Like most denizens of fence-line neighborhoods, they are financially trapped by the disappearance of jobs as local industry flees in the wake of the toxic-exposure lawsuits and revelations. Moreover, their houses won't sell. "No one wants to live on a toxic waste site," summarized Anniston activist David Baker.

Their bank is even refusing to let Dunbar homeowners Tambitha Blanks and her husband refinance their house because arsenic showed up in the groundwater as recently as 2012. The appraiser told her the poisoned groundwater affects the value and ability to refinance for anyone within 3,000 feet of the site. "If you know there's toxic stuff near the property that you're looking at, you're not going to buy it," Blanks said. "That's just evident. What do you do? We're kind of stuck in a hard place right now."

Residents of heavily African American and Hispanic communities like Dunbar; Anniston, Alabama; and East Chicago, Indiana, have grown up breathing noxious air. Many suffer physical damage including cancer, diabetes, and neurological impairment that lands them in wheelchairs or nursing homes. But even banned toxins can devastate communities, as the people of Triana, Alabama, know.

Triana, Alabama: This Is Your Brain on Pesticides

Visionary biologist Rachel Carson published *Silent Spring* in 1962. At once rigorous, powerful, and poetic, Carson's warning of a future in which nature stands in ruins, depleted beyond her ability to renew herself, struck a deep national chord.

Carson warned specifically of overusing chemicals like the pesticide DDT, which she blamed for the waning of species like the double-crested cormorant, the herring gull, and even America's iconic bald eagle.

The book drew critical vitriol from pesticide makers and their scientists, but Carson had portrayed the deadly persistence of human-made poisons in a manner that stoked the heart of America's environmental conscience. DDT was banned in the book's aftermath, around 1970, and for most of the nation, *Silent Spring* was a disquieting wake-up call.

But for the people of Triana, Alabama—86 percent of them black[76]—it was already too late.

Today, Triana is a town of five hundred on Alabama's northern border near Huntsville. And not only is it black, it is also poor, with a median annual income under $10,000. But in the past, the Huntsville River softened the town's privation. Commercial fisherman Donald Malone recalls that he made seven hundred dollars a week from fishing in the 1970s,[77] and many other townspeople supplemented their income by selling fish. More importantly, most of the 1,178 people who then lived in Triana grew their own food and fished to fill their pantries.

However, the fish were dying in large numbers and in the 1980s tests found the river to be tainted by high levels of DDT. Fish taken from the waterway harbored DDT in amounts as high as 200 parts per million (ppm)—forty times the federal limit.[78] DDT was still with them, and it is still with us today because it does not break down naturally in the environment. It has persisted in the food chain concentrated in animals, including edible fish.

"I was born and raised on the river," recalled Malone. "We made our living off it, and that's been taken away from every commercial fisherman."[79] Eating the fish from the river is now out of the question, and so is gardening for subsistence: DDT persists in the ground, rendering the food grown in contaminated soil poisonous.

In fact, much of the community's rampant disease today is ascribed to DDT exposure even though the pesticide was banned decades ago, in 1970.

Because the Army owned Redstone Arsenal, the local facility where DDT had been manufactured, the EPA ordered it to clean up the river and test residents of Triana for DDT contamination. But the Army refused, pointing fingers at the nearby Olin plant. The EPA asked the Justice Department to force the Army to clean up the DDT, but the justices in turn washed their hands of the matter, denying that the department had any power to force the fractious agencies to comply.

Meanwhile, the Centers for Disease Control studied the debility and illness in Triana's remaining five hundred residents and found staggeringly high levels of DDT and polychlorinated biphenyls (PCBs), which are known to increase the risk of heart disease, stroke, and kidney problems. The bodies of Triana residents harbored thrice the DDT levels found in poisoned workers at DDT plants, but none of the tested Triana folk had ever worked in such plants.

Although this major health threat to residents of Triana was discovered in 1978, the federal government did not act until five years later, after the mayor of Triana filed a class-action lawsuit in 1980.

One eighty-five-year-old resident, Felix Wynn, had 3,300 parts per billion of DDT in his blood, more DDT than has ever been found in any other human being.[80]

In 1980, Triana residents settled out of court with Olin for $19 million, $6.8 million of which paid legal fees, leaving each resident only about $2,000 annually for five years—money that is now long gone. The remaining $5 million went to address health care.[81]

DDT was so widely used that most Americans are still exposed to it.[82] High blood levels of DDT or its metabolites are associated with neurodevelopmental problems in children.[83] Because it persists in the soil, food, environment—and in our brains—Rutgers University scientists were able to measure DDT in their subjects' brains and correlate it with some disease patterns.

They found that people with Alzheimer's have levels of DDT and its metabolites that are four times higher than their peers without Alzheimer's. However, the sick had more than high levels of DDT: they also shared a genetic predisposition, suggesting that interaction between DDT and the genes may be needed to develop the disease. Alzheimer's would not be the first pesticide–neurological disease link: pesticides have already been strongly implicated in Parkinson's disease.

Other neurological diseases are also the product of interactions between genetic susceptibility and the environment, making it very difficult for epidemiology to identify clear associations between an exposure and a disease, especially in nations like the United States, where a broad range of genetic susceptibility reigns.

Organophosphate pesticides, such as the chlorpyrifos used widely in agriculture, on golf courses, and for mosquito control under the names Dursban and Lorsban, are neurotoxic. Prenatal exposure can lead to structural abnormalities of the brain, according to several studies. On November 6, 2018, this insecticide was approved in the United States and the EU "on the basis of a toxicity test that has now been found to be faulty," writes Philippe Grandjean on his website, where the specific testing flaws are detailed.[84] He adds that its residues cling to fruits and "appear in the urine of children, even those living in countries that don't use the product."

The United States had planned to gradually eliminate its use, but the Trump administration's EPA canceled this banishment even though, notes Grandjean, "a federal appeals court ordered the EPA to ban the pesticide due to the risks to brain development seen in studies of children."[85]

Tetrachloroethylene (also called TCE or perchlorethylene) is a widely used solvent in dry cleaning, paint, spot removers, and suede protectors. According to *Neurological Teratology*, early childhood exposure[86] carries an increased risk of neurological and psychiatric problems.

David C. Bellinger, professor of neurology at Harvard Medical School and professor in the Department of Environmental Health at the Harvard T. H. Chan School of Public Health, calculates that pesticides alone cause a cumulative national IQ loss of 16.9 million points, and the largest portion of that loss—5.7 million points—comes from prenatal exposure.

This conversion of brain damage to IQ points gives us only a rough understanding of the damage, as Philippe Grandjean points out in *Only One Chance*, because chemicals vary in their effects on the brain. Some don't affect general intelligence, but instead erode specific elements of cognition.[87] Methylmercury hampers memory; lead shortens the attention span; pesticides distort spatial perception. The polybrominated diphenyl ethers (PBDEs) used to render children's clothing and furniture flame-resistant, are also linked to cognitive and behavioral performance in school-age children.[88]

Quite a few other chemicals common in fence-line neighborhoods have been shown to poison the brain, including azides, carbon monoxide, cyanides, decaborane, diborane, fluorides, hydrogen phosphide, hydrogen sulfide, pentaborane, phosphine, and phosphorus.[89]

Toxic Reservations: Poisoned Earth, Troubled Waters, and Lowered IQs

In 2016 a national spotlight fell on the Dakota Access pipeline, and the Standing Rock Sioux tribe soon became the most visible victim of the Trump administration's disastrous changes in environmental policy. When he reversed the Obama administration's decision to deny a permit to drill beneath the Missouri River, some wondered whether the $500,000 to $1 million Trump had invested in the pipeline provided motivation. Although a spokesperson claimed Trump had sold his shares, Kelcy Warren, chief executive of the pipeline's builder, Energy Transfer Partners (ETP), had donated $100,000 to Trump's presidential campaign.[90]

Trump appointed Scott Pruitt to head the Environmental Protection Agency despite his well-known hostility to the agency's agenda. Until his July 6, 2018, departure, protections for beleaguered fence-line communities seemed to be neglected in favor of multibillion-dollar oil and gas companies. From Pruitt's status as a climate-change denier to his determination to eviscerate environmental protections, the EPA's new direction overshadows the government's actions pertaining to Standing Rock.

ETP planned to complete the approximately twelve-hundred-mile-long $3.7 billion pipeline in order to carry 470,000 barrels of crude oil daily across four states. The pipeline will swell profit margins for oil companies, but the Standing Rock Sioux point out that it will also contaminate drinking water and desecrate sacred burial sites. They refuse to accept the pipeline's construction, and their statement of resistance reads in part, "Americans know this pipeline was unfairly rerouted towards our nation and without our consent."

Native Americans and their supporters, including the environmental activist group Greenpeace, have gathered in North Dakota camps to hold sacred ceremonies[91] and to protect the Missouri River, the only water source for the Standing Rock tribe. The news media have taken note with regular updates that bring unwonted attention to Native Americans threatened with environmental hazards.

Trump's former law firm filed a complaint on behalf of ETP that characterized Greenpeace and other Sioux Nation supporters as "wolf packs" of corrupt NGOs. It deployed the RICO Act, typically used to facilitate organized crime prosecutions, against them. But an ACLU "friend of the court" brief argued that "the First Amendment prohibits companies from suing their critics out of existence."[92]

Even before the pipeline's completion, Sioux fears materialized as it ruptured, spilling 84 gallons of oil in Tulare, South Dakota, south of the resistance camps.[93]

Eagle Mine

Standing Rock is one of many skirmishes that have broken out between indigenous nations and the U.S. government over industrial wastes dumped on Native American reservations.

Some tribal governments have accepted waste storage, even nuclear waste, or mining for the millions of dollars of income they bring to a demographic that suffers twice the poverty rate of the United States as a whole. Only external pressure, including pressure from the National Congress of American Indians, prevented the Skull Valley band of Utah's Goshute tribe from committing their land for the storage of spent nuclear fuel. The tribe's home is already surrounded by a chemical weapons depot, a military test site, and a facility for the production of magnesium.[94]

But the primary "beneficiaries" of coal mining and power plants on indigenous lands often are not the native tribes. In the case of the Black Mesa region of Arizona, indigenous home of the Diné (Navajo) and Hopi peoples, four of five people living on the affected Navajo site do not have running water: their water aquifer has been tapped to supply the former coal slurry pipeline. Moreover, only half of those living on the Navajo and Hopi reservations have electricity, despite the fact that the power transmission lines cross the reservations to deliver electricity to the southwestern United States and California.

These communities rely upon natural resources for survival and hold reverence for the earth and good stewardship of these resources as cultural pillars.

This cultural mandate is threatened as Native American reservations have become the preferred sites of uranium and coal mines, leading to polluted waterways and tribal lands.

One is the Eagle Mine, the nation's only primarily nickel-and-copper mine, which is located on Michigan's Yellow Dog Plains and owned by Lundin Mining Corporation. It began production in late 2014, and is expected to generate 360 million pounds of nickel, 295 million pounds of copper, and small amounts of platinum, palladium, silver, gold, and cobalt by 2022. A coalition that includes the Keweenaw Bay Indian Community and the National Wildlife Federation appealed the issuance of its mining permit and groundwater discharge documentation based on concerns about water contamination. Their fears have a compelling basis. The Eagle Mine uses the sulfide mining method, which extracts metals from sulfide ores. When these ores are crushed, the sulfides are exposed to air and water, catalyzing a reaction that produces highly caustic and toxic sulfuric acid.

The acid drains into nearby waterways and groundwater, a phenomenon called acid mine drainage.

When water sources become acidified, plants, fish, and other wildlife that have provided food for centuries are poisoned, and the people lose not only potable water but the fruits of their treaty rights for hunting, fishing, and gathering.

But the most direct environmental threat to cognition and IQ is posed by coal-fired plants, which release neurotoxic methylmercury. This is the form of mercury that most often causes brain and spinal cord damage, that reduces IQ and causes mental retardation as well as permanent motor dysfunction.

We've known this for a long time. England's industrial revolution heightened workers' exposure to mercury's cognitive dangers, one of which was so familiar that it made its way into Victorian children's literature. *Alice's Adventures in Wonderland,* a still-popular escapist fantasy, features the Mad Hatter, whose condition was a genuine feature of British industrial life. English haberdashers used mercury to process wool into felt for hats, and hatmakers who inhaled its volatile vapors suffered brain damage, memory loss, tremors, and loss of intelligence. This was compounded by psychological changes like irritability, low self-confidence, depression, apathy, and shyness.[95] These signs and symptoms marked a disease that physicians called *erethism mercurialis* and the public called Mad Hatter disease.[96]

Methylmercury is especially damaging to the developing brains of fetuses and young children, depending upon the amount and time of exposure. Most people are exposed by the consumption of mercury-contaminated seafood, like that which caused the devastating, decades-long outbreak in Minamata city, Japan.

The epidemic was caused by the release of methylmercury in the industrial wastewater from the Chisso Corporation's chemical factory from 1932 to 1968. Methylmercury bioaccumulated in shellfish and fish in Minamata Bay and the Shiranui Sea, and was eaten by residents for thirty-six years while the government did nothing. People with Minamata disease suffered a movement disorder called ataxia, hand and feet numbness, muscle weakness, the loss of peripheral vision, and damaged hearing and speech. But the neurological damage could also be extreme, including insanity, paralysis, coma, and even death within weeks. Minamata disease also offers an example of the complex interplay between genetics and environmental poisoning because a congenital form of the disease affects fetuses in the womb.

In the United States, minority groups are most heavily affected by mercury poisoning.[97] A few examples of the coal-fired plants in and near impoverished Native American reservations include the following:[98]

The Four Corners Steam Plant,[99] one of the largest coal-fired generating stations in the United States, is located on Navajo land in Fruitland, New Mexico.

The Peabody Western Coal Company and the Desert Rock[100] coal-fired plant are just two of the many coal-fired plants and strip-mining operations in the Black Mesa region, with approximately 21 billion tons of coal and a value of $100 billion.[101]

The Absaloka Mine[102] of southeastern Montana was extended into 3,660 acres of the neighboring Crow reservation in 2008.

Colstrip Steam Plant, Montana's largest coal-fired power plant, sits on lands of the Northern Cheyenne tribe, surrounded by five large strip mines.

In 2012, the Associated Press analyzed EPA data and found that 10 percent of all U.S. power plants operate within twenty miles of reservation land. Moreover, many of these fifty-one energy-generating centers are more than half a century old and operate without protections for the fifty reservations they abut. Moreover, the EPA is considering reducing these meager protections. In February 2019, its acting administrator Andrew Wheeler indicated he will take steps to undo the Mercury and Air Toxics Standards (MATS)[103] regulations that limit the mercury and other toxic effluents that plants are allowed to release into the air.[104]

Although most Native Americans live outside reservations, the wealth of coal-fired plants in and near impoverished Native American reservations with little or no access to health care preferentially assails the intelligence and IQ of this marginalized ethnic group.[105]

Unlike African Americans and Hispanics, Native Americans constitute a relatively small ethnic group—only 2 percent of the U.S. population[106]—but we should remember that it is small because of centuries of genocide.

Nonetheless, Native American lands are home to coal mining and coal plants that disproportionately subject indigenous people to the brain-damaging environmental hazards of the coal industry.[107]

In 2006, NYU School of Medicine professor Leonardo Trasande and a team of other environmental health scientists calculated how much health damage can be attributed to mercury emissions from coal-fired power plants by analyzing mental retardation associated with methylmercury in all U.S. babies born in 2000. They then calculated how much is attributable to coal-fired power plants.[108]

Their results? All human-generated exposures of methylmercury cause a lowering of IQ that results in 1,566 additional cases of mental retardation every year. This represents 3.2 percent of new cases of mental retardation in the United States, which has cost the nation $2 billion annually.

Mercury emissions specifically from U.S. power plants cause 231 cases of mental retardation cases annually. In other words, one in every two hundred cases of mental retardation in America is caused by emissions from power plants. These cases alone cost the country $289 million every year. But the real cost of mercury emitted from coal-fired power plants is its injury to the brains of American children, particularly children on the reservation, whose risk is greatly magnified by their proximity to these toxic sites.[109]

Protecting the brains of these children entails far more than preventing exposure to emissions at home or school. Protection must begin in the womb, as the next chapter explains.

CHAPTER 4

Prenatal Policies: Protecting the Developing Brain

Even when she is not at work, Shirley Baker, the Anniston, Alabama, nurse whom we met in Chapter 3, devotes her time to improving the welfare of her poisoned neighbors. She had worked for Mothers and Daughters Protecting Childhood Health before she joined her husband in Community Against Pollution, a group he founded. Together they confront the EPA and industry to fight for her neighbors' health care and for the cleanup of their town's chemical morass. "You have to fight, you have to cover the ground you stand on, as my grandmama used to say."[1]

But Baker is mired in a private battle as well, because her medical background serves to heighten some anxieties about living in Anniston's hot zone. She breastfed her daughter because she wanted to give her baby the best possible start in life, conferring breast milk's many nutritional and immunological benefits. As a nurse, she knew that breastfed babies tend to be healthier, and even smarter, than other children.

But in a cruel irony, Baker now worries that Anniston's toxins may have concentrated in her breast milk, and that nursing might have escalated her daughter's exposure, jump-starting her behavior problems and catapulting her into the special education class in which she now struggles to learn.

It is not uncommon for parents in polluted communities of

color to face similar dilemmas. Obeying public health urgings to breastfeed and provide a diet rich in fruits, vegetables, and "brain food" like fish is usually the best way to give one's children a healthy physical and intellectual start.

But what happens when toxic chemicals war with perinatal advice, making such mundane-sounding health mandates ambiguous or even dangerous?

Food for Thought?

Parents find other medical advice even trickier to follow. Doctors' recommendations to give babies an early diet rich in fruits and vegetables, for example, seem like unassailable guidance. But they can expose children to brain-eroding pollutants lurking in unexpected places.

After birth, black and Hispanic toddlers suffer the highest rates of exposure from soil and dust, including that found in lead-tainted housing.[2] But for most American formula-fed babies, the greatest lead and arsenic exposure risk emanates from an astonishing source—commercial baby food.

When the Environmental Defense Fund analyzed eleven years of federal data, it detected lead in 2,164 baby-food samples—from grape and apple juice to carrots, teething biscuits, and sweet potatoes. One of every four apple and grape juice samples exceeded federal lead limits of 5 ppb.[3]

Consumer Reports found that even one in ten samples of "organic" juices harbored more arsenic levels than federal law permits. Other tests detected lead in 20 percent of baby food samples—that's one in every five.[4]

Formula-fed infants get most of their lead from water, but that changes as they age. Food is the greatest source of exposure for two of every three toddlers.[5] Black children's exposure to lead is much higher than other ethnic groups' and comes from their immediate environment—water, housing, dust, and local industry. But African American and Hispanic children are exposed to lead-imbued food hazards *in addition to* their excess residential contamination.

The Environmental Protection Agency estimated that over 5 percent of U.S. children ingest more than 6 micrograms per day of lead, an amount that exceeds 1963's federal lead limits*—though we now know that no level of lead intake is safe for children.

How does all this lead end up in baby food, of all places? Scientists blame the widespread use of lead arsenate insecticides, which are now banned, but which linger in the soil for decades, poisoning the vegetables grown in it and tainting the meat of animals fed with them.

Such poisoning, including from pesticides that linger in soil and water long after they have been banned, is so widespread that nine out of every ten Americans harbor pesticides or their byproducts in their bodies. Even types of produce that we think of as "healthy," such as spinach and strawberries, are also widely tainted by pesticides. For example, six out of ten samples of kale are contaminated with Dacthal, or DCPA, which the EPA labeled a human carcinogen in 1995. But others, such as avocados and sweet corn, are relatively free of such pollution according to the Environmental Working Group's "2019 Shopper's Guide to Pesticides in Produce."[6]

"Avoiding all sources of exposure of lead poisoning is incredibly important," said Dr. Aparna Bole, pediatrician at University Hospitals Rainbow Babies and Children's Hospital in Cleveland. "But the last thing I would want is parents to restrict their child's diet or limit their intake of healthy food groups."

Thus, parents who strive to provide the recommended fruits and vegetables to their children face an impossible choice: the healthy food that promotes brain-building also exposes their infants' brains to lead and arsenic.

---

* Six micrograms is the maximum daily intake level set by the Food and Drug Administration in 1993.

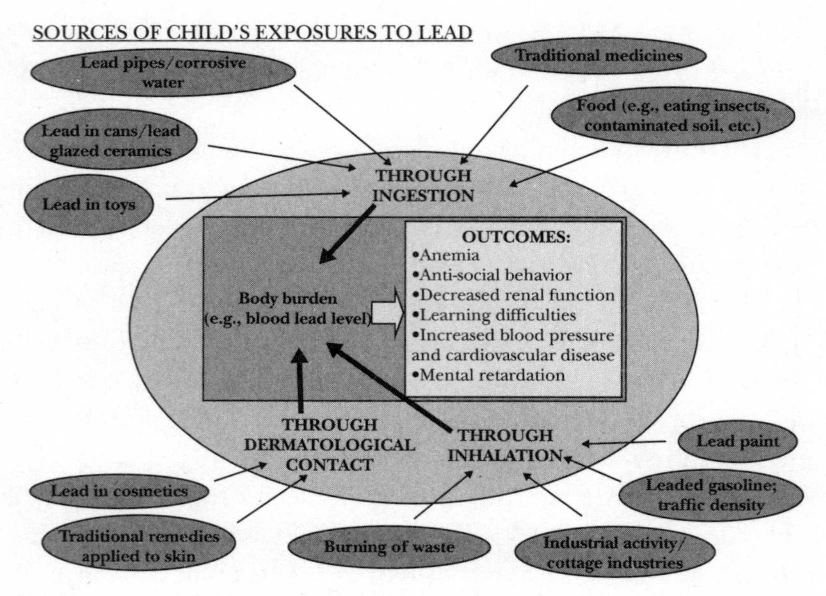

Lead poisons children via a breathtaking variety of routes, from industrial air pollutants to food, paint, and gas.

This is not the only manner in which parents have to choose between optimal nutrition and keeping their babies' brains safe.

Breastfeeding is a very important solution to this exposure, although it is not a universal one. Usually, new mothers can avoid or at least mitigate some brain-threatening exposures by breastfeeding. However, as Shirley Baker learned to her horror, other toxic agents are present in, or even concentrated in, breast milk, raising the question of how to safely give babies the benefits of breastfeeding.

A woman's doctor should be able to advise her about this during prenatal counseling and well-child care. And fortunately, Ruth A. Lawrence, M.D., in her classic *Breastfeeding: A Guide for the Medical Profession,*[7] shows exactly how to avoid hazards while making an unassailable case for the mental and physical benefits of breastfeeding.

Lawrence, a professor in the Department of Pediatrics and Neonatology at the University of Rochester and the medical director of the poison center where I once worked, also details the relative popularity of different modes of breastfeeding in U.S. ethnic groups. Hispanic mothers breastfeed at the highest rate of any U.S. ethnic groups; African Americans breastfeed at a rate lower than both Hispanics and whites, but they have shown the largest increase over time. Asian rates vary greatly by ethnic subgroup, and statistics on Native American rates are too sparse to be reliable.

The fact that breastfeeding allows infants to escape at least some of the lead and arsenic exposures suffered by formula-fed babies may help to explain why exclusive breastfeeding is associated with higher verbal intelligence and thicker parietal cortices (the regions of the brain governing functions such as sensation, vision, reading, and speech) as well as better cognitive performance.[8] But fish in a mother's diet is also key, according to a 2006 paper published in the *British Medical Journal.* The research showed that omega-3 fatty acids in breast milk, known to be essential constituents of brain tissues,[9] could at least partially account for an increase in the IQ of offspring.

A 2013 study published in the *International Journal of Epidemiology* also found that when they reached adolescence, the children who were breastfed performed better on full IQ tests, which measure verbal IQ and performance IQ.[10]

Beyond Genetics

If your brain works well enough to read a book, drive a car, and hold a job, thank your parents,[11] and not just because they contributed its genetic scaffolding. Approximately 83 percent of the brain's development takes place within the last three months of pregnancy and the first two years of life,[12] so you owe your well-functioning brain to their early care, including constant daily attention to your diet, medications, and environmental exposures—before and after you were born.

Contrary to conventional wisdom, genetics alone does not create a functional intellectual capacity, explains Professor of Education Girma Berhanu of the University of Gothenburg, Sweden. "The fetus could have the genetic potential of a gifted child, but if the potential is not enhanced through proper nutrition and medical care, there is a possibility that the child's development could be severely retarded."[13]

The infant brain is composed of more neurons than an adult's, not fewer. Like the process of painstakingly creating a sculpture from a featureless block of marble, refining the brain involves an early and precise process of "pruning"—the culling of superfluous neural connections. Furthermore, a baby's cortical centers of sensation and higher thought are actually more richly connected, with more excitatory synaptic action, than an adult's.

This helps babies to assimilate vast amounts of far-ranging information with ease. We lose this gestalt as we age but it lingers long enough to allow an infant to amass a prodigious amount of information and skills in a short time.[14] Within a few years an infant progresses from a sleeping bundle to a toddler that walks, talks, and learns ten new words and asks dozens of questions a day—*if* her brain is properly fed, nurtured, and protected from harm, including environmental insults.

Thus, your cognitive future depends on the perinatal environment you are provided. Copious research reveals that many early toxic exposures and deprivations can be disastrous for the brains of the very young. If a child's brain development is hobbled by poisons in his early environment, he may never catch up.

But poor parents often lack the resources necessary to enrich and protect their children's health. As Barbara Ehrenreich observed in *Nickel and Dimed,* "In poverty, as in certain propositions in physics, starting conditions are everything. There are no secret economies that nourish the poor; on the contrary, there are a host of special costs."[15] And parents of color are more likely than others to be poor.

Some of the starting conditions and special costs that preferentially target poor families of color are biological, including the poisoning caused by built environments that remove much of the control over a baby's earliest exposures from the hands of her parents.

For example, industrial pollution pervades housing, schools, water, and food—even baby food—threatening the brains and IQs of children from before their birth. And environmental and public health scientists have documented how the creation of food swamps, the targeted marketing of alcohol and tobacco products, and the siting of poison-spewing industries in poor areas of color all distort or short-circuit early brain development and thus intelligence, lowering IQ.

As we read in Chapter 1, the Flynn effect has documented a U.S. IQ rise of approximately three points per decade, Yet cognitive prospects now appear less rosy for the young, who are losing ground.[16] As international studies like the Trends in International Mathematics and Science Study (TIMSS) and the Programme for International Student Assessment (PISA) suggest (as do the falling U.S. National Assessment of Educational Progress and SAT scores), U.S. students perform poorly, with less than a third (32 percent) of U.S. students having attained proficiency levels in mathematics in 2009. By comparison, half of Canadians and 63 percent of Singaporeans demonstrated such proficiency.[17]

The causes are likely to be multifactorial and to include poorly performing schools and our immigration policies. Although many express concern that allowing immigration from countries with lower levels of educational achievement[18] leads to poorer cognitive performance in the United States, our laws exacerbate the problem when we deter immigration by the highly skilled from the "wrong countries."[19]

But missing from most policy discussions of lost cognitive

power among our young is a key factor: research now shows and quantifies early damage to their vulnerable brains by toxic heavy metals, industrial chemicals, and air pollution, damage that may begin in the womb, distorting behavior as well as cognition.

Even very low levels of exposure can wreak cognitive havoc.[20]

Young and Defenseless: The Vulnerable Brain of a Child

Chapter 3 notes that the young brain is exquisitely sensitive to chemical assault.

Why? Especially when it comes to poisoning, a child is not just a Mini-Me. Children differ dramatically from adults in their vulnerability to environmental poisons, including how toxic substances impair their brains and the manner in which their bodies metabolize chemicals.

From the moment a fertilized egg is implanted in the womb and throughout the first few years of life, a child's nervous system experiences prodigious growth differentiation and development. Structures are formed and critical connections are established on a precise, unforgiving timetable. Toxic exposures can lower intelligence and distort behavior by destroying these brain structures or by preventing or distorting the necessary connections within them.

The exposures responsible for this damage can be frank poisons, endocrine disruptors (which interfere with hormonal signals that direct fetal growth), or chemicals of unknown function. Because the child's developing body has neither mature immune protections against exposure nor the ability to repair itself, the damage may be irreversible and lead to loss of intelligence.[21]

The precise timetable of fetal development involves what Philippe Grandjean calls "critical windows of vulnerability." A wealth of studies document how birth defects occur in concert with key developmental events. If a mother is exposed to a neurotoxin during a critical window, its effect on her child's brain and thinking may be disastrous, although if the exposure happens a day later there may be no measurable effect. Developmental steps occur at specific times, and in a particular sequence at specific locations. As Philip Landrigan and his team write, "Implantation of the egg occurs on gestational day 6 to 7; organs begin forming on days 21 through 56; the neural plate forms between days 18 and 20; arm buds appear on days 29 to 30, and leg buds follow shortly after on days 31 to 32; testes differentiation occurs on day 43, and the palate closes between days 56 and 58." Similarly, there are critical phases of brain development, which overlap in a complex and intricately coordinated manner, with each phase offering an opportunity for chemically induced disruption and damage if it occurs at the critical time.

During the last trimester of pregnancy, brain cells are formed at a rate of about two hundred cells per second. The new cells must move as far as 1,000 times their own length to find their exact positions in the brain and nervous system.

Every cell transmits electrical signals through its newly formed extension, or axon, and there is plenty of room for error: if they were placed end-to-end, the axons from a single budding brain would encircle the globe four times. These nerve cells communicate via electrical impulses and neurotransmitters across junctions called synapses, which function like on or off electrical switches. During the child's first month of life, it spends most of its energy building its brain, including the creation of 1,000 synapses every second.[22]

The infant's brain-building phases include

- making brain cells (neurulation and neurogenesis),
- moving cells to their proper location (cell migration),
- growing axons and dendrites to link nerve cells (neuronal differentiation and pathfinding),
- developing synapses or junctures of communication with other cells (synaptogenesis),
- refining or pruning the synapses (naturation), and

- forming the "insulating" tissue that surrounds nerve cells and enables rapid, efficient communication among them (gliogenesis or myelination).[23]

Early exposures to chemicals like lead and phthalates (dibutyl phthalate and bis [2-ethylhexyl] phthalate) can interfere with these tasks, harming the brain in ways that become apparent only in later life. (As a nasty grace note, many chemicals that harm the budding brain harm the developing reproductive system as well.)

Many insist that a physiologic filter called the blood-brain barrier is effective enough to bar such poisons from assaulting and harming the developing brain. But the BBB is imperfect. It helps to protect the brains of adults from harmful exposures, but it has not been fully formed in infants. Moreover, some common environmental chemicals, like arsenic, can damage it beyond functionality, even after it matures. As a result, some exposures have been demonstrated to be three to ten times more toxic to children than to adults. In other cases, a chemical that may not harm the adult brain at all can cause devastating injury to a child.[24]

Not only are children more sensitive to many chemicals that damage the brain, they are exposed to higher doses, pound for pound, than adults. One part per billion (ppb) of a chemical like benzene ingested from water, air, or food causes greater exposure to a child than an adult for multiple reasons:

- A child younger than six months old drinks seven times more water per pound of body weight than the average U.S. adult.
- Children between one and five years old eat three to four times more per pound of body weight than the average adult.
- Infants at rest breathe twice the volume of air, pound for pound, than resting adults.[25]

- Children two years old or younger have twice the relative body surface as an adult. Because absorption through the skin is a common route of many poisons, such as aromatic hydrocarbons, this multiplies the youngest's exposure.
- Children explore the world by putting things in their mouths, and a bad taste doesn't necessarily discourage them from doing so. Just doing what children normally do can increase doses of chemicals. Hand-to mouth transmission while crawling near the source of many toxins on the floor and ground exposes them to chemicals that adults may evade.[26]

For specific environmental poisons, the differential effects on children can be dramatic. In Chapter 2, I explained that the same dose of lead that causes lifetime IQ deficits in two-year-olds produces no effect in adults. Many other common environmental exposures are much more dangerous for children than adults. For example:

Nitrate. Prolonged exposure to tap water with 20 ppm (parts per million) nitrate can kill an infant but has no observable effect on an adult.

Mercury. Exposure in the womb at 100 ppb (parts per billion)—that's equivalent to one drop in 118 bathtubs—significantly increases learning deficits, while an adult exposed to that same dose suffers no measurable effect.

Radiation. Children exposed to radiation have a much higher incidence of cancer than adults exposed to the same dose.

PCBs. Levels of fetal PCB exposure that cause learning deficits that persist through adolescence cause no measurable effects on adults.[27]

In Chapter 2, I documented the neurological devastation wreaked by heavy metals like lead, which alone drains each U.S. birth cohort of 23 million IQ points annually. But other prenatal threats, like poor nutrition, sap cognitive power. So does tainted air, as well as exposures to PCBs, phthalates, pesticides, pathogens, endocrine disruptors, alcohol, tobacco, and other toxic industrial chemicals.

Some of these risk factors may be unfamiliar, but they are key to understanding the exquisite vulnerability of the young brain. Endocrine disruptors (ED), for example, change the development of the fetus in the womb by interfering with thyroxine and other hormonal signals that direct fetal growth.

ED chemicals harm brains by mimicking natural hormones. And since the body can't distinguish these chemicals from natural hormones, it responds to the stimulus, often with disastrous consequences. Exposure to ED chemicals that mimic growth hormones, for instance, can result in gigantism. Alternatively, exposure to ED chemicals can trigger physical responses at inopportune times. Certain chemicals, for example, trigger insulin production when it is not needed, which can cause ketosis. Other endocrine disruptors block the effects of a hormone to excite or inhibit the endocrine system and cause insulin overproduction or underproduction.

There are many endocrine disruptors, but the most infamous example is diethylstilbestrol, or DES. This synthetic hormone was prescribed as a medical treatment for conditions such as breast and prostate cancers.[28] As Robert Meyers's riveting book *D.E.S: The Bitter Pill* recounts, doctors gave DES to pregnant patients between 1940 and 1971 in hopes of reducing the risk of pregnancy complications and miscarriage.[29]

Instead, DES crossed the placenta during pregnancy, and girls and women whose mothers took the drug are forty times more likely to develop a rare vaginal cancer called clear-cell carcinoma. The Food and Drug Administration subsequently withdrew its approval, and later studies showed that DES can cause a myriad of other medical disorders, including reproductive disease and infertility in both daughters and sons.[30]

Moreover, DES shows an especially insidious aspect of some

toxic agents—the ability to damage the exposed person's genes in a manner that is passed on to her children. This illustrates, yet again, the intersectionality of environmental and genetic risk factors.

Making matters worse, the effects of some endocrine disruptors like DES—including diminished intelligence, lowered fertility, and aberrant behavior—are not detected for years or even decades after the exposure. Studies also link endocrine disruptors to attention deficit hyperactivity disorder and the burgeoning of autism spectrum disorders, as well as other brain abnormalities that translate into lowered intelligence—and lower IQ scores.[31]

The endocrine disruptor is just one class of what Harvard environmental health professor Philippe Grandjean calls "brain drainers." These are chemical thieves of cognition that Grandjean, who is the head of the Environmental Medicine Research Unit at the University of Southern Denmark as well as a professor at the Harvard T. H. Chan School of Public Health, says lower intellectual potential in exposed people, especially children. (This book's Appendix includes a complete list of these chemicals.)

It's likely that other prenatal brain-affecting chemicals exist but have not yet been recognized as such. But those we are aware of already pose a staggering threat to the brains and behavior profiles of exposed U.S. children, especially children of color, who are most often thrust into the proximity of brain drainers, even before birth.

Some insist that these chemicals are not dangerous or that they are present in concentrations too low to pose a health risk. Often, critics point to a lack of evidence that these substances are harmful. But absence of evidence sometimes reflects not harmlessness but a research vacuum. You'll remember that unlike the European Union, where the precautionary principle reigns and chemicals used near humans must first be tested for safety, most U.S. chemicals are not investigated for their effects on human health until after an injury is suspected. So those who claim the chemicals are safe because no evidence of their injuries exists are wrong: until proper tests are performed, we simply cannot know their effects. Moreover, researchers and environmentalists speak of the difficulty in obtaining funding to investigate some exposures, an obstacle that is especially difficult to overcome in cases when exposure hinges on racial identity rather than economics.

Unfortunately, even after U.S. tests are performed, misunderstandings concerning toxicity—and even mythologies concerning poisoning—are rife. This means that the environmental harms to young children and fetuses have been dramatically underappreciated. For one thing, an exposure adjudged "safe" because it presents in amounts far too small to affect an adult can transform an infant's brain development, his mental and behavioral functioning, and the course of his life.

Heavy-Metal Mortality: Vanished Children

In Chapter 2, I described how PCBs as well as common heavy metals and metalloids like mercury, arsenic, and lead wreak intellectual devastation on children of color, especially African Americans. But they do more than stultify existing children: they also kill unborn ones.

A 2017 report exposed "hundreds of excess deaths" of fetuses in Flint, Michigan, between 2013 and 2015 after the city switched to lead-poisoned Flint River water. Health economists Daniel Grossman and David Slusky found that between 218 and 276 more children should have been born, and that these "missing children" succumbed to fetal death and miscarriages caused by waterborne lead exposure[32]—a crisis that left so many dead that the city's 2014 fertility rate plummeted.

The water-purity change was restricted to a specific period, allowing clear comparisons of Flint's fertility and fetal health rates before and after the switch, when fetuses were exposed to tainted water in utero for at least one trimester. Because Flint

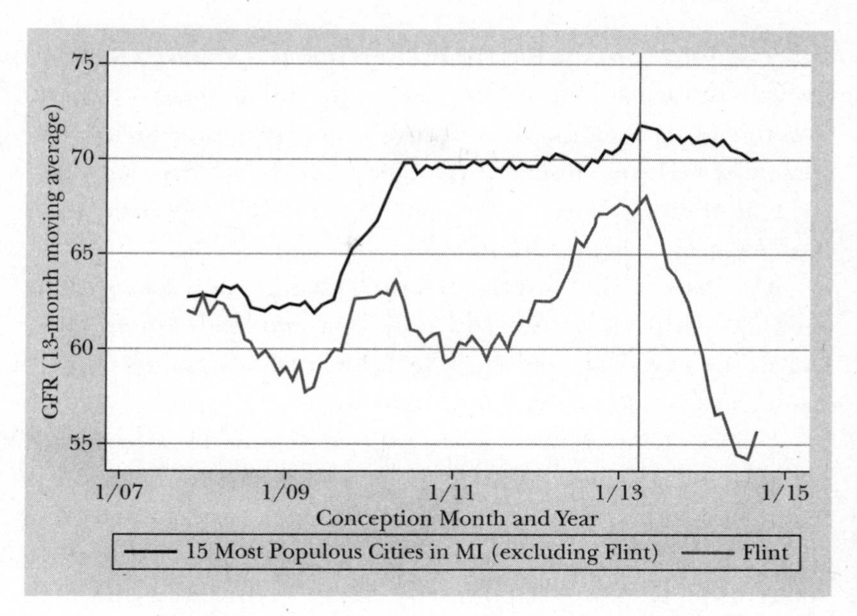

Unlike in other Michigan cities, the birthrate in Flint fell sharply, reflecting the many fetuses killed by the city's environmental exposures, notably its lead-tainted water. Note: *The red vertical line is at April 2013, which is the last conception date for which no affected birth rates are included in the moving average.*

was the only city in the area that switched its water supply, studies could meaningfully compare data with surrounding cities. No other Michigan cities recorded such a drop in fertility.

Even so, this count of missing babies is probably significantly underestimated because the investigation included only fetal deaths reported within hospitals, and did not include abortions or miscarriages that occurred before twenty weeks' gestation.[33]

Flint was not alone. In 2013, the economists also found lead-driven fetal deaths rose as much as 42 percent, and birth rates fell in Washington, D.C., during the years 2007 and 2008, the same period during which the city endured its own lead crisis as levels rose in its drinking water. Lead levels in the District of Columbia had also peaked in 2001, then fell again in 2004 when public health correctives were put in place to protect pregnant women.[34]

The children born during the periods of high pollution suffered growth abnormalities, an increase in the prematurity rate, and lowered birth weights. By studying which neighborhoods were most affected, researchers were able to correlate the density of lead plumbing to both the markers of maternal blood lead levels and lead poisoning in children.[35]

The babies that survived didn't emerge unaffected. The publicity about the nation's racially stratified lead-tainted water makes the news that lead sabotages the brains of infants of color dismaying, but not completely surprising.

Doctors tell parents to avoid exposing their children to lead when possible: a clear strategy, but one that is hard to follow when lead's presence in housing is hidden, as it was to the residents in East Chicago, Indiana, or when parents are actually steered to lead-tainted housing by health agencies, as they were in Baltimore.

Air Pollution

In Chapter 3, we saw how polluted air's elevated carbon dioxide (CO_2) levels trigger the hypoxia of asthma, which as studies show, leads to lower intelligence. Airborne heavy metals like lead and mercury as well as hydrocarbons, such as those in fuel exhausts and industrial emissions, also cause rampant brain damage to adults and children alike. These conditions compromise both the cognitive development and the function of a brain by depriving it of sufficient oxygen.

When it comes to air pollution, the developing brains of children fare the worst. Their greater lung surface area relative to their body size means that they suffer a greater relative exposure to noxious gases and suspended particles than adults.

For pregnant women and their fetuses, the damage is even more profound. Air pollution attacks fetal brains when they are at their most sensitive. Even small insults during crucial developmental windows can translate into fateful cognitive problems that persist for a lifetime.

How can air pollution doom the life—or the intellectual development—of a fetus? A May 2017 *Time* article, "Preterm Births Linked to Air Pollution Cost Billions in the U.S.," gave the answer, but under the wrong headline: more important than the monetary losses is the fact that air pollution causes 16,000 premature births in the United States each year.[36] We don't know how many of these premature babies suffer lifelong mental disability, but we do know that pollutants derail the manner in which a woman normally delivers air to her fetus. Airborne pollutants also disrupt the endocrine system, which prevents the production of an important protein that regulates the pregnancy.

The air inside homes also threatens the brains of very young children. Carbon dioxide concentrates in poorly ventilated homes where it can kill sleeping families. By reducing oxygen intake, it also slowly erodes the brains of survivors, causing a laundry list of symptoms, including cognitive decline.

Vermin

We don't usually think of microbes and other living things as causing pathological behavior or lowering intelligence, but we should. *Infectious Madness: The Surprising Science of How We "Catch" Mental Illness* documents the many ways in which infection—sometimes carried by vectors and microbes—causes mental disorders and disordered behavior. So does "The Well Curve," a 2015 article in *The American Scholar* that discusses microbes' important role in lowering intelligence. Chapter 5 of this book discusses in detail the cognitive costs of microbes that preferentially afflict people of color.

We have seen that mold, dust mites, and cockroaches have been shown to drive up African American and Hispanic asthma rates at a cost to intelligence that has yet to be quantified.

However, despite a paucity of attention, microbes also pose a hypertension threat.[37] Hypertension is most often caused by genetics, diet, excess weight, stress, and a sedentary lifestyle. But it can also be triggered by... rodents. Mice and rats can harbor Seoul virus, a type of hantavirus that, among other signs and symptoms, raises blood pressure.

In the mid-1990s, a high-profile Centers for Disease Control study investigated a deadly hantavirus outbreak among Native American populations in the Four Corners region, the symptoms of which were chiefly respiratory. The outbreak, which eventually spread to the general population, was found to be carried by infected mice.

Four years earlier, Gregory Gurri Glass of the Johns Hopkins School of Public Health had reported that infection with different, rat-borne hantaviruses—Seoul virus—in "inner-city" populations is associated with an increased occurrence of hypertension.[38] His team discovered that people living in poor urban neighborhoods infested with rats more often have viral antibodies indicating they have been infected by the virus, and connected undetected hantavirus infection to the high rates of hypertension among African Americans.

"Emerging diseases don't just happen in Zaire, Kenya or Rwanda," Gurri Glass told me at the time. "They are happening throughout this country now." But a few years later he said that progress on the Seoul-hantavirus-hypertension question had stagnated because neither he nor New York City researchers on a parallel track had obtained funding to study the connection more extensively.

"The general view seems to be these infectious agents are not of particular relevance," he lamented, noting that the U.S. Army ended funding of pertinent research. "We're competing for resources with other societal ills.... I'm just hoping we can continue."

But how is this virus-mediated hypertension pertinent to our discussion of intelligence? Hypertension is a well-known risk factor for physical problems like heart disease, and stroke. However, like asthma, it also sabotages intelligence: in addition to causing stroke, which often diminishes mental function, memory, and intelligence, hypertension erodes intelligence more directly. High blood pressure in otherwise healthy adults between the ages of eighteen and eighty-three is associated with a measurable decline in cognitive function, according to a University of Maine report in the journal *Hypertension*. Constantly elevated pressure weakens blood vessels including those in the brain, which can harm the brain by compromising oxygen delivery or heightening the risk of small brain injuries. Cognitive decline is a documented long-term effect of hypertension, even in children.[39]

Even worse, a 2012 report in *Neurology* suggests that hypertensive mothers may actually hand down their hypertension-induced cognitive decline to their children.

A Finnish study found that men whose mothers' pregnancies were complicated by hypertensive disorders scored almost five points lower on tests of cognitive ability than men whose mothers did not have high blood pressure during pregnancy.[40] "Our study suggests that even declines in thinking abilities in old age could have originated during the prenatal period when the majority of the development of brain structure and function occurs," wrote Katri Räikönen, Ph.D., of the University of Helsinki.[41]

What's more, one in ten pregnancies are complicated by hypertensive disorders like preeclampsia, which is more common and severe in African American women than whites, and is linked to premature birth and low birth weight, factors that are independently associated with lower cognitive ability.[42]

Parents seeking a solution to vermin and pollution exposure that could threaten cognition face an uphill road. Moving away from heavily polluted areas and out of dilapidated, vermin-infected housing is an obvious step, but can be an unaffordable one for low-income families, or even middle-class ones, due to mortgage redlining and stubborn U.S. housing segregation. However, even moving away from bus depots, manufacturing plants, and dry cleaners, which are sources of the volatile cleaning solvent trichloroethylene (TCE),[43] provides limited protection. So does investing in HEPA vacuum cleaners,[44] which control semi-volatile organic compounds such as phthalates, flame-retardants, pesticides, polycyclic aromatic hydrocarbons (PAH), and polychlorinated biphenyls (PCBs) that otherwise persist in the air and dust.[45]

You can further reduce indoor pollution by using air conditioners and maintaining their filters while keeping the windows closed whenever possible, and certainly during high-traffic periods.[46]

If you own your home, keeping it vermin-free with the aid of professional exterminators is an expense you cannot afford to forego. If you rent, know your rights as a tenant: most cities have an agency like Legal Aid or a housing authority that offers free legal representation to tenants who cannot afford a lawyer. Enlist its help in your quest for safe, healthy housing. It may help you to move or to force your landlord to comply with health codes. You can discover the pertinent laws in your area by checking the Nolo site Renters' & Tenants' Rights at https://www.nolo.com/legal-encyclopedia/renters-rights.

Municipal codes typically prohibit renting infested residential property and fine landlords for failing to maintain premises free of vermin. If this is the case in your city, you may be able to use the law as leverage to pressure the landlord to hire professional exterminators.

Waterborne Microbes

Waterborne poisons and microbes also contribute to fetal death and brain damage in communities of color and elsewhere.[47]

Lead is not Baltimore's only intelligence-threatening pollutant. Its waters harbor microbes such as bacteria and copious amounts of suspect or disease-causing algae, like *Pfiesteria piscicida.*[48] Moreover, the risk is much higher in some areas: residents of the Chesapeake Bay's watershed are exposed to bacteria from sewage at more than 2,000 times the healthy levels. After storms, levels of fecal coliform bacteria from raw sewage and animal waste are extremely high in Baltimore County's Back River. In 2001, they ranged from 640 to 2,135 times higher than the healthy levels.

The EPA and Justice Department sued Baltimore in 1997 over its malfunctioning sewage system, and municipal officials made a legal commitment to perform $900 million in repairs and upgrades.[49] Even so, the water of the Back River continues to harbor at least one oxygen-deprived "dead zone" where nothing can live. Its sewage and microbial waste is joined by a witches' brew of pollutants that are also known to affect cognition. "Our waters have a little bit of Prozac in them, a little bit of oral

Anglers' Days, Trips, and Expenditures by Population Group: 1996
(16 years of age or older; numbers in thousands)

	All Anglers	*African-American Anglers*	*Hispanic Anglers*	*Female Anglers*
Anglers	35,246	1,802	1,185	9,509
Days of Fishing	625,893	40,131	16,685	112,841
Mean Days of Fishing	18	22	14	12
Trips	506,556	32,550	13,562	94,267
Mean Fishing Trips	14	18	11	10
Total Hunting Expenditures	$20,694,946	$813,836	$695,532	$3,003,094
Trip Expenditures	$15,386,271	$583,687	$513,346	$2,334,499
Mean Trip Expenditures	$437	$324	$434	$246
Equipment Expenditures	$5,308,675	$230,149	$182,186	$668,595
Mean Equipment Expenditures	$151	$128	$154	$70

contraceptive hormone, a lot of caffeine," summarized Hopkins professor Thomas A. Burke.

Meanwhile, experts from the Johns Hopkins Bloomberg School of Public Health reassured residents that if they "limit their consumption of fish to the state's guidelines, they should be fine."

But because some areas are much more heavily affected than others, limiting exposure is not always possible. Nor is it practicable to expect this of people who fish not for sport, but for subsistence, supplementing their diet with a free source of quality protein. As David Baker, the environmental activist in Anniston wryly observed, "We're fishing for survival: we're not going to throw fish back."[50]

Mercury-Tainted Fish: To Eat, or Not to Eat?

For pregnant women and new mothers, subsistence fishing presents another danger to their children's brains: mercury.

Doctors urge pregnant women to enrich their diets, even before conception, with foods that will help build their baby's brains. But in earlier chapters we have read how environmental poisoning prevents many poor African Americans and Hispanics from gardening for fruits and vegetables and discourages fishing for high-quality protein.

In urban communities, food swamps abound, creating the same dearth of nutritious food. Ethnic enclaves without supermarkets or farmers' markets are often dubbed "food deserts." However, the term "food swamps" is more accurate because the absence of healthy food in these regions coexists with an abundance of cheap, easily available, non-nutritious, sugary, and fatty fare, as well as the targeted marketing of potent forms of alcohol and of tobacco. Communities of color are often dependent on "corner stores" and bodegas where a handful of faded vegetables can cost more than an entire fast-food meal. Parents typically must pay for taxis to reach supermarkets and return with groceries, adding to their cost.

Such dietary danger zones are tied to income, but are more

strongly tied to race. And when it comes to nutrition, infants, once again, are not just miniature adults. The large surface area of their intestines relative to their body volume often makes infants more vulnerable to toxins in food. They also metabolize foods and medications through different pathways than adults do. For example, babies younger than two years old should not eat honey, because unlike adults, their less acidic stomachs cannot neutralize the botulism spores it may contain and they may contract the fatal foodborne disease. Babies require special vitamin supplements; mothers of breastfeeding babies must adjust their diets and avoid certain medications. Prenatal counseling and nutritional advice during well-baby visits are essential in maintaining a baby's mental as well as physical health.

However, when faced with pollution dangers, even doctors and researchers can find it difficult to give good advice regarding food. As a result, recommendations concerning the advisability of pregnant women eating fish from tainted waters vary based on the purity of the water, the species of endemic fish, and sometimes, by the sophistication of public health analyses.

Take mercury for example. Methylmercury is the form of mercury that causes neurotoxicity, but the relatively benign elemental form of mercury suffuses some waters. This leads some to assume that eating fish from the latter waters is safe. But it may not be, because some waterborne microbes transform elemental mercury into the brain-toxic methyl form by adding a few atoms—a methyl group, written as CH_3. Several types of bacteria are responsible, notably sulfate-reducing bacteria and iron-reducing bacteria. There are many, but some common examples include salmonella, pseudomonas, and campylobacter, which are familiar because they cause food poisoning and other illnesses in humans. Other bacteria that carry the specific genes *hgc*A and *hgc*B can also transform elemental mercury into methylmercury as well.[51] Additionally, some industrial wastes contain forms of iron that encourage mercury methylation when discharged into water.

Scientists also warn that pregnant women who consume tainted fish may face damage from industrial residues like PCBs (polychlorinated biphenyls) as well as from waterborne mercury. So do their fetuses.

Moreover, anglers may not realize that the size of the fish they catch is a key factor in poisoning risk. Toxins like mercury bioaccumulate—that is, they become greatly concentrated when larger fish eat smaller, tainted ones. Larger fish are more toxic than smaller ones from the same waters.

Formal studies have addressed the question of whether it is better for a pregnant woman to eat fish from tainted waters or to abstain and forgo seafood's desirable brain-building nutrients. In 2004, the FDA and the EPA advised women of childbearing age to limit their seafood intake to about three servings a week during pregnancy to avoid exposure to trace amounts of methylmercury and other neurotoxins.

But this was followed by the National Institute on Alcohol Abuse and Alcoholism's 2007 study, which made a very different recommendation. Joseph Hibbeln found that the significant nutritional benefits of omega-3 fatty acids sacrificed by not eating enough fish outweigh the dangers of consuming mercury-tainted fish.[52] His study of nearly 12,000 pregnant women in Great Britain found that the equivalent of two or three servings of seafood a week resulted in more intelligent children with better developmental skills, and he wrote in *The Lancet* that even more seafood in the diet might be advisable. "Advice that limits seafood consumption might reduce the intake of nutrients necessary for optimum neurological development."

One researcher suggested to me that this is a concern for Asian Americans but not for African Americans, because there are no national data suggesting that they practice subsistence fishing. I was surprised to hear this, because I have seen it so often and known it to be common in areas where I have lived, worked, and visited. My own father and his inner-city friends often fished and sometimes hunted to help feed our family of seven.

And there is the plight of cities like Triana, whose residents practiced fishing to supplement their diets until pollutants rendered the fish dangerous to eat, as recounted in Chapter 3. But I understood that my anecdotal experience might not reflect national trends, so I looked for data.

As the skeptical researcher predicted, I did not find large national studies showing that African Americans engage in substance fishing that puts them at risk for mercury exposure. And I did find studies documenting it in Asian American communities. But this might mean that national researchers are simply not investigating the phenomenon in African Americans. "Absence of evidence is not evidence of absence," responded environmental sociologist Robert Bullard, when I approached him for a reaction.[53]

Despite the paucity of national data, some other studies, including analyses conducted in California and New York, looked at fish consumption by several racial groups and found that African Americans consume more high-mercury fish than the norm, writing, "Non-Hispanic blacks and women grouped in the 'other' racial category [whose ancestry is Asian, Native American, Pacific Islands, and from the Caribbean Islands] had significantly higher BHg (blood mercury) concentrations than did non-Hispanic whites."[54] In some studies, their consumption rivals that of Asian Americans, which is itself a highly diverse group.

Until more studies are done, we won't know whether this pattern is repeated in larger swaths of the nation. In the absence of national data, some may be tempted to dismiss African Americans' risk as unsubstantiated. But although unassailable proof is important, addressing known health risks in a timely manner is more important. We have seen what happened when a "prove it beyond every doubt before we take action" approach led to the annual loss of 23 million IQ points and $50 billion to lead.[55]

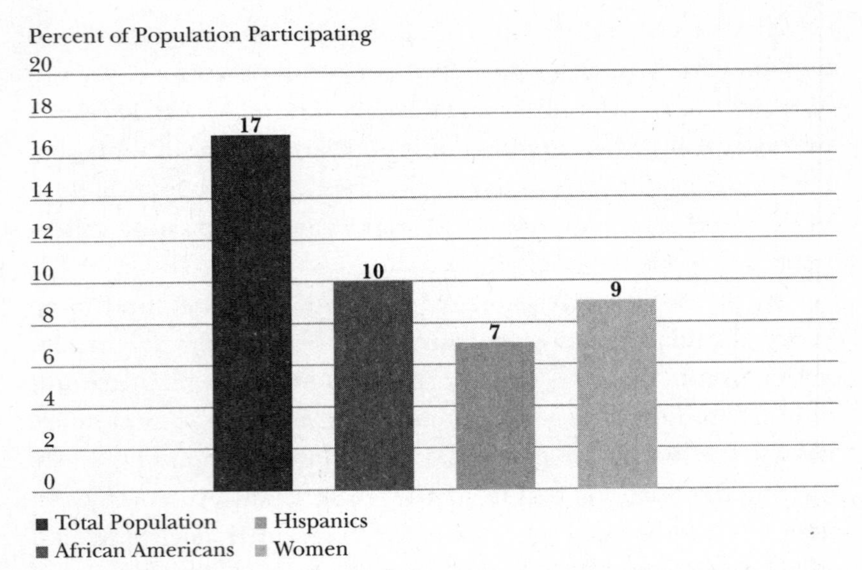

As this graph shows, significant numbers of African Americans engage in subsistence fishing to supplement their diets, as do Hispanics.

The stakes are so high that the precautionary principle applies here. We must address the risks of waterborne mercury.

Fortunately, the EPA's inspector general released a report in April 2017 castigating it for failing to adequately protect vulnerable populations, including African Americans, against contaminated fish.[56] This acknowledgment is a necessary first step in addressing hidden risk.

Until national data that clearly indict or exonerate mercury exposure are collected and analyzed, one solution is to take a more individual approach to detecting mercury in prospective mothers. In 2017, a Danish study determined that testing the hair of a pregnant woman for mercury on the day of her first ultrasound can help reduce fetal exposure,[57] because it permits her doctors to inform her if her baby is at risk, monitor and treat her appropriately, and provide an incentive for avoiding additional mercury. Doctors should consider a hair test for minority mothers, and any woman who is worried about mercury risk should request such testing.

Parents themselves can undertake another important solution by choosing varieties of fish that are known to be low in mercury. The Food and Drug Administration lists shark, swordfish, king mackerel, and tilefish as the four fish with the highest levels of mercury—avoid them. Instead choose light tuna (avoid white albacore tuna, which has a higher mercury content), salmon, pollock, catfish, shrimp, and mackerel (the smaller mackerel fish, which will have lower concentrations of mercury). You should also avoid fish that are salted, as pollock often is.

Tilapia is inexpensive, rich in vitamins, and low in mercury, but many avoid it because of misinformation circulating on the Internet. There is no scientific evidence that "bacon is better" for you than tilapia, as the Internet alarmists claim: a single study's conclusions were taken out of context. More information is available by searching factcheck.org.[58]

The FDA further recommends limiting yourself to about twelve ounces of fish a week, the amount contained in two average meals, but pregnant women should seek guidance on this from their physicians.

Prenatal Peril

The mercury exposure/omega-3 fatty acids risk-benefit ratio is just one aspect of quality prenatal nutrition related to intelligence.

The poor nutrition so common in urban food swamps also potentiates other causes of lowered intelligence, notably lead poisoning. But a high-nutrient diet that includes varied fruits and vegetables offers some protection against the ravages of lead poisoning because a wealth of nutrients, like the calcium in canned salmon, sardines, leafy green vegetables, and milk, makes it harder for lead to be absorbed and strengthens the immune system. Iron, found in beans, lentils, lean red meat, raisins, prunes, and other dried fruits also blocks lead absorption.[59]

In addition, pregnant women, new mothers, and their children are too often deficient in amino acids, iodine, calcium, iron, phosphorus, and vitamins such as folic acid, which prevents cognition-sapping neural-tube defects. A varied, balanced diet and supplements will further protect the infant brain.

Although all developing fetuses and children need similar nutrients and similar protections against poisons, race and ethnicity sometimes complicate acquiring them. We are rarely speaking of physiological differences, although there are a few, such as the increased need of dark-skinned people, like some African Americans, Asians, and Hispanics, for additional vitamin D supplements. (Vitamin D deficiency is more prevalent among African Americans, even young, healthy ones, in part because their darker skin blocks sunlight's UV rays, which are necessary for our bodies to manufacture vitamin D, and are relatively scarce at North America's higher latitudes.)[60]

Most racially disproportionate nutritional risks are created by environmental and social pressures. Unfortunately, however, doctors do not always include environmental advice when counseling the pregnant women and mothers who most need it. In one study, monitoring found about one hundred different toxic pollutants including lead, mercury, toluene, perchlorate, bisphenol A, flame retardants, perfluorinated compounds, organochlorine pesticides, and phthalates in pregnant women, with forty-three such chemicals found in every woman tested. Yet only 19 percent of these women's doctors had discussed pesticides and only 12 percent discussed air pollution, even though their dangers are well established. Only four out of every ten doctors said they routinely discussed such contamination with pregnant women.[61]

Only 8 percent of physicians warned their patients about bisphenol A (BPA) and 5 percent advised avoiding phthalates. Nine percent of the doctors—fewer than one in ten—told their patients about the brain-draining polychlorinated biphenyls (PCBs) often found in fish.[62]

Why the silence? Many doctors say their priority is to protect pregnant women from more immediate dangers, and that warning them about environmental risks may create undue anxiety. Although Dr. Naomi Stotland of San Francisco General Hospital knows that her low-income patients on California's Medicaid program are probably at higher risk of toxic exposures, she told *Scientific American* that she didn't discuss environmental health with them for a long time. Why? "The social circumstances are so burdensome. Some colleagues think the patients are already worried about paying rent, getting deported or their partner being incarcerated."[63]

Thus, unborn children of color are awash in environmental hazards about which their doctors remain silent.[64]

The situation is even worse for women who do not receive the recommended level of prenatal care. Women without access to prenatal care often know far too little about nutritional risks to their unborn children. Fewer than one fertile woman in three knows, for example, that taking prenatal folic acid supplements reduces birth defects and eliminates spina bifida.

Cultural risk factors threaten the infant brain as well. Some folk remedies such as *greta* and *azarcon* teas, taken for indigestion, contain lead, and so do some candies and delicacies from countries like Mexico and China that are imported and sold cheaply in dollar stores or bodegas to people of all ethnicities. *Nzu* and *poto,* nostrums for morning sickness, can contain dangerous levels of lead as well.

Then there are the various nonfood items that are consumed, sometimes in large quantities, by those with pica. Pregnant women are among those who sometimes satisfy cravings for chalk, starch, or even mud, all of which can harbor dangerous pollutants and microbes. "Calabash chalk" a natural material composed of fossilized seashells, is consumed by pregnant women in some West African cultures and in parts of the United States as a treatment for nausea. It is found in North American ethnic stores as well, but it can contain high levels of lead and other risky adulterants.[65]

Cleaning House

To minimize the foodborne poisons that threaten your fetus or baby, eliminate as many as possible from your household. In order to reduce exposures to bisphenol A (BPA), avoid buying food and beverage cans with resin liners. This is especially important in light of a New York City study that measured prenatal and postnatal BPA exposure until age five in inner-city African American children and found that children with the highest BPA exposure levels showed more problems on tests that measured emotional reactivity and aggressive behavior.[66]

Read the labels of your plastic containers before buying them to be sure they do not contain chemicals called phthalates. In fact, it is a good idea to avoid buying any foods packaged in plastic and to avoid processed foods as much as possible. Food co-ops, farmers' markets, and stores that carry organic fare may offer the option of paper containers or allow you to bring your own.

Few people have the time to cook all their meals from scratch, but many people find making the week's entrees in advance and freezing them until use allows them to control sugar, fat, and additives, while saving time in the long run.

Cosmetics contain suspect chemicals, too. Read the label of cosmetics and avoid using those with phthalates, as well as lipsticks that contain lead: look for additives that contain "plumb" in their name. Cheaper foods and candies purchased at "dollar stores" sometimes come from countries where food can contain unacceptable levels of lead, so this may not be the best place to save money: pooling funds with friends to buy high-volume cases of domestic foods and treats at big-box stores may be safer.

It is important to clean in order to minimize vermin and tracked-in pollution, but many cleaners and pesticides carry their own host of toxic problems, so try using baking soda, vinegar, and nontoxic commercial cleaners instead of toxic products during pregnancy.[67]

Fetal Alcohol: A Hidden Scourge?

In June 1998, two weeks after completing a yearlong fellowship at Stanford University, Serena walked into a bodega on 147th Street, a block from her new apartment in pre-gentrification Harlem. Repulsed by the fumes from young smokers as she waited her turn in the dingy, claustrophobic storefront, she watched a young man buy a forty-ounce bottle of malt liquor without being asked to produce ID. Another requested "two loosies" and was surreptitiously handed a couple of cigarettes in exchange for a dollar.

Intimidated by the sketchy characters loitering about, she tried not to stare at the clerk and wondered whether it was safe to say anything. But as she looked away, a lurid poster just beneath the counter riveted her attention. Could she be seeing correctly? The graphic ad featured a color photograph of two rhinoceroses copulating over a slogan proclaiming the virtues of an obscure malt liquor brand. She was shocked into silence until she heard giggling and saw two young boys pointing at it. Then she became angry. Adults might not notice the poster, but it was low enough to be directly in the line of sight of the children. "What is this and why is it here?" she demanded. "Don't you know that these children can see it?" A shrug and a smirk accompanied the clerk's bored response: "I don't know, lady; I'm not the owner.... Are you going to buy something?"

"Every block in the ghetto has a church and a liquor store," Dr. Walter Cooper, retired Eastman Kodak executive and NYS Regent emeritus, once told me. Perhaps it's inevitable that growing up in an environment saturated with enticements to dangerous addictions would lead to African Americans' suffering the nation's highest rate of smoking and second highest rate of fetal alcohol syndrome.

Over a decade after Serena's experience, even with gentrification in full swing, lurid alcohol shops continued to haunt Harlem neighborhoods. In 2011, even the historically black neighborhood of Mount Morris Park, home to $2 million brownstones and $3 million apartments, became the site of a nuance-free liquor store replete with a roll-down steel gate, Plexiglas customer barriers, and lurid neon sign decried by its appalled neighbors as "ghetto."[68] They complained to the *New York Times* that they knew a liquor store was coming, but they had expected a tony wine shop, not a bulletproof dive hawking Night Train.

What's for sale in such establishments? Something you're hard-pressed to find elsewhere: cans and bottles of malt liquor share the refrigerated case with Budweiser and Beck's, but they deliver the alcoholic wallop of a bottle of wine, not a can of beer. Fortified wines radiating colors not found in nature taste like candied gasoline because they're tailored to the tastes of the young, including underage drinkers. By the mid-1960s, liquor companies began marketing their high-alcohol malt liquor to African Americans.[69]

Sexualized posters and labels sell misogyny with the alcohol, as Ice Cube urges, "Get your girl in the mood quicker, make your jimmy thicker," and old-school Billy Dee Williams winks and says, "Works every time."

Although some middle-class white youths also partake, they usually must visit the 'hood to procure these drinks, some made by vintners that would be familiar to their parents.

MD 20/20, so potent that it is referred to as "Mad Dog" on the street, is produced by the Mogen David Wine Company, better known for its kosher dessert wines. Night Train Express, 17.5 percent alcohol, hails from the sun-dappled California vineyards of Ernest and Julio Gallo.

But you won't find these hardcore libations among the sedate table wines on their websites, nor on restaurant wine lists. They're not for discriminating palates but for those who want maximum alcohol at the lowest prices, and they are fortified not with vitamins but with extra alcohol, keeping poor communities of color steeped in liquor, on the cheap.

Such potions were invisible when I lived in Brookline, Massachusetts, the Village, and Palo Alto. But on my infrequent forays into East Palo Alto, then 81.8 percent black and Hispanic, I often saw half-nude women posing in ads hawking malt liquor and fortified "ghetto wines" just as they did in Harlem, Baltimore, and Chicago's minority neighborhoods.

For decades, these communities have been steadily saturated with come-ons for addictive products tailored to the minority market. Between 1986 and 1989, 76 percent of billboards in African American neighborhoods of Baltimore advertised alcohol and tobacco, compared to only 20 percent in white communities. In Detroit, the ratio was 56 percent to 38 percent, and in St. Louis, 62 percent to 36 percent. Similar patterns are found in New Orleans, Washington, D.C., and San Francisco.[70]

Tobacco billboards are now widely banned, but those advertising liquor remain, and they are supplemented by depictions in rap videos, smaller posters plastered everywhere, films, television shows, and paid celebrity spokespersons.

The alcohol-soaked environment of minority neighborhoods is replicated in the womb, where it endangers unborn children.

The easy availability of liquor in African American communities leads to high rates of fetal alcohol effects (FAE), a less severe condition than fetal alcohol syndrome (FAS),[71] but one that is also caused by drinking alcohol during pregnancy. Alcohol's devastating effect on African American fetuses is a problem that psychiatrist Carl Bell says "has been hidden in plain sight."[72]

In fact, says Bell, who has studied the rates of undiagnosed FAE and FAS for decades, "FAE is the largest preventable public health problem in poor African-American communities."

Alcohol is a teratogen, an infectious, pharmacological, or other biochemical agent that acts during pregnancy to produce birth defects or dysfunction. We've known since 1973 that when a pregnant woman drinks, her child may emerge with fetal alcohol syndrome, marked by unusual facial features and brain damage. Although the characteristic facial abnormalities of FAS are its most apparent symptoms,[73] brain scans performed on victims of FAS reveal damage to the brain's white matter and abnormal connections between the frontal occipital lobes, areas that govern executive functioning and visual processing. Accordingly, victims of the syndrome are faced with a variety of congenital defects, including diminished intelligence, low IQ, mental retardation, hyperactivity, difficulty concentrating, poor memory, learning difficulties, speech and language delays, and poor reasoning and judgment.

Children with FAS suffer delayed language and motor skills, their short-term memory is compromised, and they are often impulsive, acting with a lack of foresight about their actions'

FASD	**Delinquency**
Cognitive impairments	Poor cognition
Impulsivity	Hyperactivity
Trouble in school	Learning disabled with diagnosis of cognitive dysfunction and attention deficit hyperactivity disorder
Behavior (emotional) problems	Early antisocial behaviors

consequences. Sixty percent of adolescents with FAS are arrested or have other trouble with the law, such as shoplifting, and disruptions in school. They suffer from low birth weight and a slowed growth rate as well as heart, eye, and genitourinary malformations as well.[74]

Less severe manifestations of prenatal alcohol exposure include fetal alcohol spectrum disorder and alcohol-related neurodevelopmental disorder (ARND). People with ARND might have intellectual disabilities and problems with behavior and learning that cause them to perform poorly in school, as well as difficulties with math, memory, attention, judgment, and poor impulse control.

Many affected people cannot handle the social interaction and ordinary tasks of everyday living. As a whole, these mental and behavioral deficits set children up for failure in school and the workplace through no fault of their own.

Carl Bell, a Chicago child psychiatrist, has extensively documented the previously unsuspected extent of fetal alcohol disease in African American children. (Courtesy of Carl Bell)

The symptoms of FAS are often mistaken for other forms of behavioral disorders that are commonly ascribed to minority children. But subtleties distinguish the signs and symptoms of these disorders from FAS. Carl Bell notes that three of every four youths in Illinois' Cook County Temporary Juvenile Detention Center had difficulty with reading, math, communication, memory, explosive behavior, hyperactivity, poor attention skills, and social judgment. Most had other diagnoses, but FAS should have been considered in such cases.

Early diagnosis and therapy can dramatically increase a child's chances of success in life, but far too few FAS children are promptly diagnosed and given the support that could enable them to tame their behavior problems and fit into society.

Although the national rate of FAS is two to five percent of children, Bell and his Chicago colleagues documented a staggeringly high FASD rate of 57 percent in a Chicago clinic that served African American patients.[75] Their symptoms and medical history were consistent with the spectrum of fetal alcohol disorders.

It's entirely possible that FAS alone contributes significantly to the lower average IQ of U.S. African Americans. However, we require larger studies of FAS rates among African Americans nationwide to know for sure.

How could this FAS epidemic among African Americans fall below the public health radar, especially when, as Johns Hopkins University professor Ellen Silberberg points out, "The African American alcoholism rate is lower than the national average"?

Because several factors tend to veil alcohol-related brain damage among African American children:

Most African American mothers of babies suffering from FAS/FASD are not alcoholics. Health workers suspect and often screen for FASD in the babies of alcoholics, but not necessarily in the babies of women who are moderate drinkers. Moreover, even exposure to alcohol only during the third trimester of pregnancy can trigger FASD. But this is precisely when some mothers erroneously assume that their baby is fully formed and that a glass of wine or an occasional cocktail is safe. There is also a higher prevalence of fetal alcohol syndrome, although African American women are less likely to drink than whites.[76]

Mothers of FAS/FASD babies are young and tend not to know they are pregnant until two months have passed. During this time they may have engaged in social drinking that puts their baby at risk.

Health workers focus on Native American mothers. The high known rate of FAS in Native Americans leads to stricter scrutiny of their infants, but not of African American babies. Although screening for FAS is important, we must guard against the *firewater myth,* the belief that biological or genetic differences make American Indians, Alaska Natives, and other First Nations people more susceptible to the effects of alcohol, including alcohol problems such as FAS. Aside from being plain wrong, the futility inherent in this myth is a disservice to Native Americans, a study by University of Alaska psychologists found, because "believing that one is vulnerable to problems with alcohol may have negative effects on expectancies and drinking behavior... and have negative effects on attempts to moderate drinking."[77]

No law requires that U.S. newborns be screened for FAS. This means that many infants are never diagnosed, and Bell documents that as they age, their intelligence loss and behavior problems are ascribed to other causes.

Immediate protective measures have not been instituted. We swab newborn babies' eyes with silver nitrate to protect against possible damage from gonorrhea during birth. Most mothers don't have gonorrhea, but this offers safe, cheap, and effective insurance against blindness for the minority whose mothers do. We need similar protective steps for babies exposed to alcohol in utero.

Bell thinks that such a simple, safe protective treatment

exists. Based on evidence that it protects against FAS, he and others recommend that we explore the option of giving women choline, a vitamin-like nutrient that has been shown to be protective against FAS in his and others' studies. If proven effective and safe in wider studies, he urges that we adopt postnatal choline, that it be added to prenatal vitamins, and that its current dosage in conventional multivitamins should be increased.

Some studies also offer evidence that choline may protect against another brain thief, Alzheimer's, which African Americans suffer from at twice the national rate. Researchers at the University of Colorado at Denver propose that a form of choline may help prevent the development of autism, ADHD, and schizophrenia as well.[78] Choline seems to be a nutrient that demands further investigation.

Although the consequences of FAS are dire, its victims are not doomed, because protective factors and early intervention can ward off signature life failures like dropping out of school, chronic unemployment, and jail. With the proper diagnosis and support, people with FAS can enjoy success.

Just ask Morgan Fawcett, the young Tlingit Alaska Native who founded One Heart Creations to raise awareness for FAS and who travels the nation giving concerts and benefits. Fawcett is also a flutist who was honored as a Champion of Change in a 2011 White House ceremony. Initially, he recalls,

> I had no clue that I was struggling because parts of my brain didn't develop. I was struggling because my body didn't develop right.... My mother drank during pregnancy. But just because she drank during pregnancy doesn't make her a bad person. There's a stigma placed on mothers. When it comes to fetal alcohol there are only victims, never perpetrators. Even with my disabilities because I have support we are doing things differently, I've become profoundly successful. I'm a

> 4.0 student in college because we were able to adapt my learning style to my coursework.[79]

The disastrous effects of FAS can be mitigated and even prevented, says Luther K. Robinson, Professor of Pediatrics at the State University of New York School of Medicine and Biomedical Sciences. He lists the factors that protect against failure as including

- an early diagnosis of FAS (before six years of age),
- receiving developmental disability services,
- a positive, stable, nurturing home for at least three-quarters of one's life, and
- not being subjected to violence.[80]

All of these measures protect against school dropout, jail, and chronic unemployment, says Robinson.[81] Kathryn (Kay) Kelly, Project Director of the University of Washington's FASD Legal Issues Resource Center, agrees.

A Smoking Gun

Tobacco use results in approximately 434,000 deaths and costs the United States $52 billion annually in medical care.[82] Most smokers begin in adolescence: eighty to ninety percent of smokers start before the age of twenty. And according to one study, as many as half of Native American adolescents smoke.[83] Minority communities are targeted by the makers of tobacco products, a subset of which are cynically marketed to women.

Since the 1920s, women have been targeted by tobacco-marketing strategies promoting "liberated" women, such as Virginia Slims, Silva Thins, and Eve ads, the latter of which are specifically targeted to African American, Hispanic, and Asian women. In 1960, about 10 percent of all cigarette advertisements appeared in women's magazines: by 1985, this advertising rate had more than tripled.

The reassuring outdoorsy images of fit women engaged in healthy activities like hiking, jogging, and yoga belied the health hazards of smoking. So did the misleading descriptions of the cigarettes' tar and nicotine content.[84] The duplicity did not end with the advertisement: articles in magazines that ran tobacco advertising were much less likely to mention smoking as a risk factor for disease or otherwise criticize the habit than were magazines that did not run such advertising.[85]

Tobacco companies also wooed racial minority groups by papering urban communities with billboards advertising brands of tobacco and alcohol tailored to the ethnic markets. For example, the cynically named "X" brand, widely believed to refer to Malcolm X, featured a pack that opened from the bottom because company research revealed this as a favored practice of black men.

Tobacco companies also curried favor among African Americans by hiring many black executives at a time when other industries shunned them, and through large donations to cultural groups.[86]

Smoking rates are higher among African Americans than whites or Hispanic Americans, and this is bad news for their children. Tobacco is known to harm the developing baby in a myriad of ways, including prematurity and low birth weight, which themselves are risk factors for many serious disorders, sudden infant death syndrome, cleft lip/palate, stillbirth, and infant death. Smoking is implicated in other conditions, from heart defects and miscarriage to attention deficit hyperactivity disorder.[87]

This means that smoking cessation is key in protecting the prenatal brain and intelligence.

Accordingly, a 2012 study reassured us that 68 percent of

doctors warn their pregnant patients to stop smoking and to avoid secondhand smoke.[88]

Yet this statistic means that about three of ten doctors offer pregnant woman no such counseling, and these are likely to be women of color. By 2000, both the *American Journal of Public Health* and the *Journal of the National Medical Association* had warned that pregnant black and Hispanic women were the least likely to be offered smoking cessation counseling from their physicians.[89] But the most recent studies show that doctors most often fail to warn Native American mothers away from tobacco use.

This silence is stunning, especially given the important role of tobacco in many Native American cultures, which may make it harder for women to quit.

Prenatal counseling is an essential source of advice and support for women who seek to give their babies the best start in life. Tragically, women of color sometimes find censure and abuse, not support. Medical counseling and intervention have sometimes been used to vilify women of color, who, beginning in the earliest days of the United States, were blamed for poor birth outcomes by being cast as murderously indifferent mothers. High infant mortality was laid not to the privation and abuse of enslavement and, later, to segregation that separated women of color from care, but rather to "overlaying," a term doctors used to accuse black mothers of killing infants by rolling over on them as they slept. This diagnosis was reserved for enslaved mothers.

Some still harbor this biased mentality. In Milwaukee, an alderwoman recently pushed to criminalize parents whose babies died after sleeping with them if the parents had been intoxicated.

Centuries after enslavement, the African American infant mortality rate remains high—in fact, it is worse today than it was during enslavement, notes City College journalism professor

Linda Villarosa in her brilliant *New York Times* investigation "Why America's Black Mothers and Babies Are in a Life-or-Death Crisis." Current data reveal that American black infants are twice as likely to die as white ones—11.3 per 1,000 black babies, compared with 4.9 for whites.[90]

As one peruses the medical and social literature that seeks to explain the reasons, a question recurs. Simply stated, it is: "What are black mothers doing wrong?"[91]

Blame the Victim, Redux: Shades of Guilt

Expectant mothers of color all too often find public condemnation, betrayal, and jail time—or just as dangerously, silence—regarding the risks to their children's developing brains. The seminal work of Dorothy Roberts, especially her book *Killing the Black Body,* documents how racism escalates the vulnerability of poor mothers of color, who suffer stricter scrutiny from the legal system and from child-welfare agencies than do their white peers. It has not helped that the current mantra of "personal responsibility" has eclipsed the earlier public health model of corporate responsibility, and this has escalated stigmatization by public health policies.

Today, a kinder, gentler, and far more appropriate medical model dominates discussion of the newly minted "opioid crisis" and other forms of drug abuse. But a harsher legal paradigm has long focused on people of color, with a punitive animus. This punitive approach has leveraged accusations of drug abuse (real or imagined) to constrict minority women's reproductive freedom—without protecting at-risk children of color.

Take the case of twenty-seven-year-old Darlene Johnson of California, whose judge, Howard Broadman, made her an offer she could not refuse in 1991.[92] The African American mother of four had "spanked" her six-year-old daughter with a belt and an electric cord for smoking a cigarette, and her four-year-old got the same punishment when Johnson caught her inserting a wire hanger into an electrical outlet. Now Johnson, who was eight months pregnant, was facing seven years in prison for three counts of felony child abuse. Broadman offered to sentence her to just one year in jail and three years' probation—but only if she agreed to an implantation of Norplant, a surgically inserted contraceptive that can be removed only by a physician and lasts for five years.

Johnson asked whether it was safe, and Broadman assured her that it was, so she agreed. But when she discovered that Norplant is contraindicated in women like herself who suffer from diabetes and hypertension, she changed her mind. Although the ACLU and even the district attorney (!) joined her lawyers in urging Broadman to release her from the agreement, he would not.

Even Sheldon Segal, the embryologist who developed Norplant, told the *New York Times* that he was troubled by the medicated device being forced on women. "I just don't believe in restricting human rights, especially reproductive rights and I'm also bothered because this is a prescription drug, with certain side effects and certain groups of women for whom it may not be appropriate. How does the judge know if the woman is diabetic, or has some other contraindication to the drug?"[93]

The health concerns were real: Norplant was eventually removed from the market because of serious side effects including stroke, blindness, and permanent sterility.

Like many other jurists who forced Norplant on women of color, Broadman explained that he was acting with the welfare of her children in mind: by preventing Johnson from reproducing, he was ensuring that there would be no more children for her to abuse.

This reasoning is illogical: it does nothing to protect her existing children, as counseling and social-work intervention could. Moreover, preventing a child from being born is a draconian means of protecting it from spankings or abuse. Such rulings are punitive, not corrective.[94] There were many Darlene

Johnsons: 85 percent of the women on whom the legal system forced Norplant were African American or Hispanic.

Black women have also been jailed for bad birth outcomes, even imaginary ones.

In 1991, we learned that the infamous "crack baby"—always portrayed as black, born addicted, and with profound permanent brain damage caused by the mother's crack use while pregnant—was a myth. But we learned this only after the headlines blamed neglectful, drug-addled mothers for the "crack baby epidemic," further stigmatizing black women who sometimes faced legal charges for their babies' medical issues. As I wrote in *Medical Apartheid,* the "crack baby" was an imaginary golem created by shoddy, racialized research studies and journalistic sloppiness: in reality, these issues resulted from the various medical risks of poverty and poisoned environments, not maternal drug use.[95] After twenty years spent studying crack cocaine, developmental neuroscientist Pat Levitt characterized the near-ubiquitous medical despair over infant golems as "an exaggeration." "The story that science is now telling rearranges the morality of parenting and poverty, making it harder to blame problem children on problem parents," Madeline Ostrander wrote in *The New Yorker* in 2015.[96]

The mythical crack babies were actually suffering from poverty and its attendant substandard housing, exposure to violence and chaos, overcrowding, and noise, all stressors that can poison the infant brain as surely as drug abuse. Their brain damage is comparable to that of other babies born into poverty.[97]

Black women, who constitute the fastest-growing group in prisons, abuse drugs at the same rate as whites but are ten times more likely to be incarcerated for drug use while pregnant.[98] In 1994, the Center for Reproductive Law and Policy filed a complaint to the National Institutes of Health that helped to explain why. The complaint alleged that the Medical University of South

Carolina, in Charleston, had performed illegal drug testing and human experimentation on approximately forty pregnant women who had sought prenatal counseling.

The women had gone to this Charleston hospital seeking prenatal care and help kicking their drug habits, as doctors and PSAs constantly urge them to do. Instead, they were turned over to the police and arrested. Some gave birth while shackled to their hospital beds, and in many cases, the hospital also conducted illegal medical research on them. The U.S. Department of Health and Human Services civil rights director Dennis Hayashi mounted a civil rights suit because all of the prosecuted women were black, except one. The sole white arrestee, as noted in her medical record, "lived with her boyfriend who is a Negro."

This selective arrest and prosecution of black women fits an insidious national pattern. Nearly nine out of every ten women prosecuted and jailed for drug abuse while pregnant are black: the charge tends to be some variant of "delivering drugs to a minor."

Public clinics in urban settings test and report more vigorously than private clinics, and this helps drive the racial disparity.

Unfortunately, punishing mothers of color for poor birth outcomes continues today. When twenty-one-year-old Regina McKnight suffered a miscarriage in 2003, South Carolina charged her with "homicide by child abuse," alleging that drug use caused her baby's stillbirth—a claim that they were unable to prove. She was convicted anyway, and served eight years of her twelve-year sentence before the South Carolina Supreme Court overturned her conviction because her lawyer had failed to call in experts who could have refuted the state's outdated evidence against her. In 2015, Purvi Patel of Indiana was convicted of feticide and child neglect following her miscarriage and sentenced to twenty years. In July 2016, the Indiana Court of Appeals overturned her conviction for feticide but left standing her child-neglect conviction.

Arresting and jailing women for their newborns' deaths affects women of color at every socioeconomic stratum: wealthier, better-educated African American women are also likely to suffer poor birth outcomes. For example, female African American doctors and lawyers are more likely than white women without high school diplomas to suffer the death of their infants.[99] Once again, race trumps socioeconomics.

One reason for this, as Nancy Krieger reminds us, is that staracially discriminatory health policies that lower the health status of black women threaten the health of their babies as well. Krieger, a professor at the Harvard T. H. Chan School of Public Health, conducted a 2013 study of infant mortality in states with and without Jim Crow laws. She found that black infant deaths were significantly higher in Jim Crow states, but that after the passage of the 1964 Civil Rights Act, the gap between them shrank until it ceased to exist in 1970.

News of perinatal racial neglect and abuse spreads widely in communities of color and destroys the trust between women and the purveyors of prenatal care. Thus, women of color often learn to fear approaching even those institutions that render excellent, ethical support, and the brains of their babies pay the price.

Today, toxicologists warn, we are witness to decreases in IQ partnered with unexplained increases in neurodevelopmental diseases like autism spectrum disorder and attention deficit hyperactivity disorder.[100] Some, like Grandjean and Professor Barbara Demeneix, blame a battery of environmental insults by chemicals that have not been properly tested on humans. Corporations generate these, and only the law can control their emissions.[101]

However, as David Rosner and his colleagues informed us in Chapter 2, public health officials have largely replaced their strategy of forcing corporate polluters into compliance with urgings of "personal responsibility" for one's own health.

But personal responsibility has its limits. Individuals cannot control many of the exposures that directly or indirectly threaten infant health and brains, from endocrine disruptors and other noxious chemicals to air pollution, befouled water, and cockroach infestation. Moreover, some remedies, like buying potable water and nutritious brain-building foods, and obtaining quality prenatal care and supplements, are prohibitively expensive for many.

Even utilizing expensive technology has begun to come under the rubric of exercising responsible parenthood. Dorothy Roberts worries that genetic screening that can detect illness risks is not an unalloyed blessing because it is now considered an essential part of preventive medicine. Technologies that involve genetic selection—reprogenetics—shift responsibility for promoting well-being from the government to individual women by making them responsible for ensuring the genetic fitness of their children. But cash-poor women, especially black women, currently face financial and other barriers to receiving reproduction-assisting services, such as in vitro fertilization (IVF).

Solutions will lie in acknowledging this reality. Public health efforts that pressure companies who dispense environmental chemicals to remove the exposures and to fund long-overdue safety testing will serve children better than advising parents on how to minimize the damage.

Paternal Risks

We also need to better understand the ways in which fathers and male significant others affect the health of unborn children and newborns. Public health policy currently turns a blind eye to the father's role in protecting infants' brains. The equally long overdue funding of research into the role of fathers and male partners is another necessary step toward better fetal health.

Although about one in three American fathers lived apart

from their children in 2009, some, including men of color, are the primary or sole caregivers for their children. The myth of the "absent father" assumes that African American fathers tend to be uninvolved, but recent studies show that they tend to be *more* involved in their children's lives than fathers of other ethnicities.[102]

We "discovered" the intimacy between fathers of color and their children only recently, because in the past no one thought to ask. We must update shortsighted traditional patterns of research that ignore the male partner, his smoking, his interactions with his children, his age, his nurturing patterns, his drug and alcohol use, and his own exposure to chemicals and gene-altering toxins such as DES. Does drug, tobacco, or alcohol use or abuse affect sperm? Does it affect the mother's ability to stay sober? Does a smoking father harm the fetus through passive smoking by the mother? Also, we know that abusive men are most likely to beat a woman while she is pregnant: Is there a cognitive effect for the child? To what degree does a warm, supportive father contribute to a child's cognitive and behavioral health? We don't have the answers yet because virtually no one is asking; such dangerous silences are driven by sexism as well as racial bias.

The policies of the current presidential administration do not bode well for children's health, and one of the best examples of this is the rollback of Obama-era nutrition standards. In May 2017, less than a week after his appointment, United States Department of Agriculture secretary Sonny Perdue announced that he would be rescinding nutrition-enhancing policies. This now allows schools to serve offerings like high-sodium foods, fattier foods, and flavored milk with sugar and higher levels of fat.

CHAPTER 5

Bugs in the System: How Microbes Sap U.S. Intelligence

A young African American mother once asked me if these neglected infections could be responsible for seeing all those little special needs yellow buses in her urban neighborhood every morning during the school year. I responded that she just asked an amazing and profound question, but right now we have not even begun to answer it.

— Peter Hotez, M.D., Dean, National School of Tropical Medicine at Baylor College of Medicine[1]

A fetus floating languidly in utero and an infant who sleeps seventeen hours a day seem the very definition of indolence. Actually, they are hard at work as they marshal their physical resources to construct a breathtakingly complex brain.

"Being smart is the most expensive thing we do," writes Christopher Eppig, Ph.D., director of programming for the Chicago Council on Science and Technology. "Not in terms of money, but in a currency that is vital to all living things: energy."[2] Nothing is more expensive, metabolically speaking: brain-building consumes 87 percent of an infant's energy.

When a fetus or young child is infected by a parasite, bacterium, virus, or other pathogen, fighting the infection is metabolically expensive, too. While at the University of New Mexico, Professor of Evolutionary Biology Randy Thornhill and Eppig found that a child cannot do both. The energy diversion severely taxes the child's ability to produce a normal brain.

How? The brain of a fetus fighting infection, like that of a fetus exposed to chemical intoxicants (see Chapters 3 and 4), will suffer derangement or malformation that renders it incapable of certain intellectual functions. Missed developmental milestones, struggles in school, or frank learning disabilities and retardation may appear soon after birth, or as late as early adulthood. The New Mexico researchers' theory of how pathogens lower our IQs by deforming our mental development is called the "parasite-stress theory of intelligence."

Zika, Our Newest Immigrant

We are all vulnerable to the pathogens that cause this mayhem, but racial and social bias cause them to target poor people of color. These include immigrants from tropical lands but also hundreds of thousands of U.S. natives. Here, as in the developing world, infection by pathogens directly correlates with impaired cognition, memory, concentration, and lower IQ.

The most recent such pathogen to garner headlines is Zika, a virus known to attack neural tissues and sabotage brain development. It was discovered in 1947 among rhesus monkeys in Uganda's Zika forest, but by 1952 it had moved to humans. The World Health Organization explains that Zika is spread mostly by *Aedes aegypti* mosquitoes, although the virus persists for an unknown period in semen, and sexual transmission from men to women has been documented in the United States and elsewhere.[3] In fact, some think that sex has become the chief mode of transmission. Before 2007, Zika was found only in Africa and Asia, but in 2018 approximately fifteen hundred pregnant women contracted Zika, especially in New York City and in the American Southwest, where it is carried by not only *Aedes aegypti* mosquitoes but also the domestic *Aedes albopictus.*

In February 2016, the World Health Organization pronounced the burgeoning Zika virus a "public-health emergency of international concern." Only three other disasters have earned this label: the 2009 H1N1 epidemic, the 2014 Middle Eastern polio resurgence, and the 2014 Ebola epidemic.[4]

The usual symptoms of Zika disease in adults include mild fever, skin rashes, muscle and joint pain, conjunctivitis, and "malaise" that lasts for two to seven days; fetuses can suffer stillbirth, microcephaly, and eye malformations.[5] But unlike the other three disasters, the Zika outbreak threatens more than our physical health: it can also generate or encourage mental and cognitive disorders in affected fetuses.[6]

When the brains of babies with congenital defects were scanned they were found to be smaller and marked by underdeveloped structures. Complications such as Guillain-Barré syndrome and microcephaly (small heads and underdeveloped brains) in infants have risen with cases of the disease.[7]

Ominously, a 2015 Brazilian study found that most infants born to infected mothers in that country show abnormal cortices, or outer layers of their brains.

A March 2016 report in the *New England Journal of Medicine* indicted Zika as a cause of microcephaly after it analyzed the virus's genome.[8] Tatjana Avšič Županc of Slovenia's University of Ljubljana told *New Scientist,* "Microscopic examination revealed that brain cells were destroyed due to infection with the virus. While it can't be definitive proof, it may present the most compelling evidence to date that congenital brain malformations associated with Zika virus infection in pregnancy are a consequence of viral replication in the fetal brain."[9]

The infection causes the brain to develop abnormally, which can result in microcephaly, and eighty-five percent of those with microcephaly suffer significant cognitive limitations.[10] However, even Zika-infected children who escape microcephaly suffer limitations[11] that range from the minor to the significant, including speech delays, seizures, movement and balance problems, short stature, and facial anomalies.[12] Experts fear that Zika places "even infants who appear normal at birth...at higher risk for mental illnesses later in life," according to the *New York Times*. "The consequences of this go way beyond microcephaly."[13]

Conventional wisdom suggests that Zika will not be as disastrous in the United States as it has been in the developing world because of our greater density of health care practitioners and our rich disease-surveillance resources.

But the histories of other neglected tropical diseases that have made landfall in the United States suggest a darker scenario. Like HIV infection and Chagas disease, Zika may come to threaten poor ethnic enclaves of the United States that are subject to the same substandard living conditions, reduced access to health care, and environmental hazards as communities in the global South.

In fact, some scientists compare Zika's fetal damage to that seen in the children whose mothers were infected with rubella during pregnancy during a 1964–65 epidemic.[14] These children were born with deafness, blindness, and mental problems[15] ranging from brain swelling to mental retardation. Their brains pay this heavy cognitive price partly because the virus, like other infectious diseases, saps a newborn's metabolic energy.[16] Viruses like Zika also disrupt brain growth by diverting energy from their human hosts to crank out copies of themselves. Other pathogens, like worms and other parasites, infest the digestive tract, where they siphon off nutrients like iron that are necessary to build a healthy brain and nervous system. The results of such nutritional deficiencies include various mental and cognitive deficits from a reduced attention span or memory to lowered verbal intelligence. Lower IQ scores result.[17]

Moreover, Zika, like schizophrenia and bipolar disorder, also causes devastating mental disorders by directly attacking the brains of fetuses or exposing them to attack by the maternal immune system—"friendly fire" that can damage the brain instead of routing the virus, leading to a lifetime of cognitive and mental disability.[18]

Such brain damage is not as rare as was once thought. Zika's cognitive damage, for example, was originally reported to be confined to exposures during only one trimester.[19] But in March 2016, the *New England Journal of Medicine* published an ultrasound study revealing that the fetuses of 29 percent of women who had tested positive for the Zika virus were plagued by "grave outcomes." In vitro experiments show that the virus "targets and destroys" those fetal cells destined to become the brain's cortex, and fetuses are affected in all three trimesters.[20]

Guillain-Barré syndrome may also emerge in children affected by Zika, an assortment of alarming and even life-threatening symptoms that come on quickly and are caused by temporary slowing of nerve conduction. In the worst case, this dangerous loss of muscle and nervous system function can stop breathing. Recovery can take anywhere from a few weeks to a few years.[21]

So Zika is a physical disease *and* a mental disease that damages the fetal brain, sabotaging cognitive function and with it, intelligence.

Such IQ-pathogen connections can sound like science fiction, but the relationships are scientifically validated and the concept is far from new. As early as 1919, medical researchers in Queensland, Australia, probed the connection between infectious disease and intelligence, and five years later, the *Virginia Health Bulletin* asked whether its public health system was battling "Stupidity or Hookworm?"[22] as it sought to trace the connection between rampant U.S. hookworm infection and abysmal school performance in poor rural schoolchildren.

Hookworm, or *Necator americanus,* is a parasite that most often works its way into the body from the soles of unshod feet, then migrates to the small intestine where it sucks its host's blood, causing anemia, weight loss, profound fatigue, and impaired thinking. The rampant hookworm infection of the South has historically targeted the poor who could not afford shoes or decent shelter, and so fed the stereotypes of the stupid, lazy African American and the equally lethargic and slow "redneck."

But the *Virginia Health Bulletin* posed its question during the heyday of U.S. eugenics, and the prospect of an infectious, rather than a racial, driver of intelligence ran counter to eugenic theory. The infection theory specifically warred with the common belief that African Americans possess innately low intelligence. The hookworm connection failed to draw funding or the continued attention of researchers and the infestations were widely thought to have dissipated by the 1980s.

Today, however, studies use modern tools to investigate the hookworm-health connection. In Alabama's Lowndes County, 67 percent—that's more than two of every three people—are infected by hookworm. Nationwide, the National School of Tropical Medicine estimates that 12 million people are infected.[23]

We've also long known of more widespread diseases that caused sufferers to lose a profound degree of brain function. One such disease is paresis, common in the United States until the 1940s. Affected patients could no longer work, remember basic information, care for themselves, converse with others, navigate independently, or even walk. Eventually they were confined to institutions, then to beds, until the erosion of their brains prevented basic bodily functions and killed them. In the end, paresis was found to be the tertiary and final stage of syphilis infection, so this cognitive devastation was caused by a bacterium.

Paresis is now a foreign disease, but it once was a common diagnosis in early-twentieth-century U.S. asylums. When penicillin was found to cure syphilis, paresis disappeared from the West, and now persists only in the global South and other health

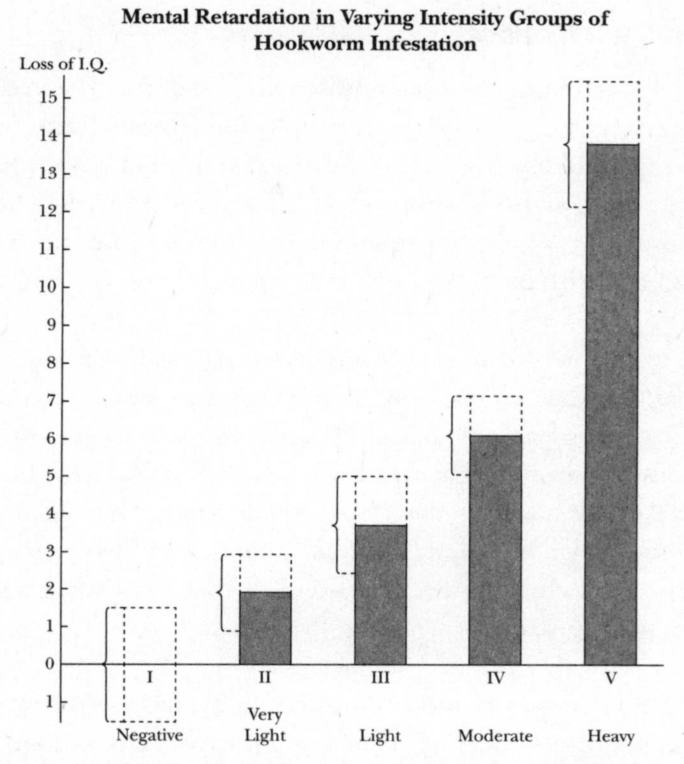

Heavy hookworm infestation correlates with the highest degree (approximately 14 points) of IQ loss; very light infestation correlates with low losses of 1–2 IQ points. Note: *Section of bar covered by brackets represents probable error.*

care vacuums where people without access to health care still suffer and die from it after losing their memory, speech, emotional stability, and cognition.

In a similar manner, unmasking pathogens that sap intelligence will allow us to staunch the loss of IQ points.

Thanks to the work of scientists like Eppig, Thornhill, and Hotez, new technological knowledge of infection-cognition connections allows us to draw connections, to quantify exposure, and to better understand the mechanisms by which pathogens destroy intelligence.

American Infections

In the United States, paresis is now a disease of the past. In many ways, Zika is a disease of the future in the United States, which has so far avoided the waves of debility that Zika has caused in other, poorer nations. But even if Zika never flourishes here, a plethora of diseases that sabotage intelligence and depress IQ are already with us.[24] As Hotez et al. write,

> The Big Five diseases—Chagas disease, cysticercosis, toxocariasis, toxoplasmosis, and trichomoniasis—are quite common here among the poor.... Diseases like schistosomiasis, a parasitic worm infection, hookworm, and Toxocariasis, which strikes the poor in the US—actually reduce intelligence. They reduce IQ among kids, and there are studies to show that when chronic infections occur in childhood [they can] reduce future wage earning by 40 percent.[25]

These diseases sound unfamiliar, like conditions from elsewhere, and quite recently, even some doctors have denied their presence in the United States. They insist that these neglected tropical diseases (NTDs), as they are still called, only affect the developing world and immigrant enclaves in this country.[26]

When the patient of a doctor at the Baylor College of Medicine received a letter saying that blood he had donated was being rejected because of a positive test for Chagas disease, the doctor exploded, "The test is wrong. That disease doesn't exist in the United States!" But it does: at least 330,000 U.S. residents, and maybe as many as one million, suffer from Chagas disease. The Big Five now affect at least 14 million U.S. residents.[27]

Chillingly, these disorders target poor enclaves of African Americans and Hispanics, sapping the brainpower of poor people of color at home, just as they do abroad.[28] Such intellect-sapping diseases target communities in subtropical areas like Houston, Alabama, and Florida. Why?

Because NTDs are not "foreign diseases" anymore. News reports portray Zika as an exceptional tropical disorder, but it is common in one important respect: its appearance in more than twenty American states[29] is part of a decades-long pattern. Anthony Fauci, director of the National Institute of Allergy and Infectious Diseases, wrote in the *New England Journal of Medicine,* "Dengue hit with a vengeance in the '90s. Then we had West Nile in 1999, chikungunya* in 2013, and lo and behold, now we have Zika in 2015 and 2016. This is a disturbing, remarkable pattern."[30]

After linking high infection rates with lowered IQ, Randy Thornhill and Chris Eppig asked *how* infection deranges the developing brain and saps intelligence.

In several ways, it seems. Parasites and worms lodged in the intestines siphon off nutrition that the brain needs to develop; "That's like the next thing after slapping the fork out of your hand. Food gets into your mouth and into your body, but then you can't absorb those nutrients," says Eppig.[31] As infections sap her energy, the infant's body must struggle to fight off the invaders *and* to properly construct her brain with inadequate resources.

But prenatal infections can also damage the brain in horribly versatile ways.

Microbes may breach the linings of our intestinal "second brain," which contains more neurotransmitters than the brain in our skulls.[32] From there, chemical messengers travel to the brain, where they cause excessive and chaotic neuronal firing and damage.[33] Of this "leaky gut" model of destruction, Thornhill's team wrote: "If exposed to diarrheal diseases during their

---

* I found no studies that ascribe cognitive symptoms or any effect on IQ to chikungunya.

first five years, individuals may experience lifelong detrimental effects to their brain development, and thus intelligence."[34]

Or, the brain may be attacked by the baby's own immature immune system in a florid but inaccurate attempt to destroy the pathogenic invaders, resulting in damage to the brain's own tissues.

Infectious agents can cause damage to adult brains as well. You may recall that 87 percent of an infant's energy goes to building a healthy, functioning brain. But a completed adult brain is also high-maintenance, requiring fully 25 percent of an adult's energy to remain in good working order, and when that energy is siphoned off to fight infection, the brain can be rendered ineffectual and in some cases irreparably damaged.

Mapping Intelligence

Average IQ varies with geography, both across and within nations according to the controversial but widely utilized assessments in *IQ and the Wealth of Nations.*[35] Despite the authors' notoriously sloppy methodologies, described in Chapter 1, the book purports to rank nations by their IQs and it is constantly recruited to support hereditarian theories.

But other scholars use the book's rankings as well.

In 2010, Chris Eppig, Corey Fincher, and Randy Thornhill compared data from it and several other IQ rankings to health data from the World Health Organization in order to draw global correlations between a country's infectious-disease burden and the average national IQ of its inhabitants. The scientists found that the greater the burden of infectious disease, the lower a nation's average IQ.[36]

But how robust is this correlation, and what of other variables that might drive intelligence? Nigel Barber, for example, writes that IQ varies in accordance with dramatic educational differences. Donald Templer and Hiroko Arikawa believe that because cold areas are difficult to live in, evolution favors higher IQ in cool climates.

Such factors may contribute to IQ, but what is the most powerful driver of intelligence? Thornhill's team's tests corrected for all of the above factors as well as genetics, nutritional levels, national wealth (gross domestic product per capita), average temperature, and for several measures of education. In order to investigate the effects of education on U.S. IQ, for example, Thornhill repeated the analysis across the United States where standardized compulsory education exists.[37] In the end, the team wrote, "Infectious disease remains the most powerful predictor of average national IQ when temperature, distance from Africa, gross domestic product per capita and several measures of education are controlled for."

Geraint Rees, director of the UCL Institute of Cognitive Neuroscience, told *The Guardian* that their conclusion appears valid: "It explains about 50 to 60% of the variability in IQ scores and appears to be independent of some other factors such as overall GDP."[38]

Eppig's team concluded that "infectious disease is a primary cause of the global variation in human intelligence."[39]

Their analysis didn't seem to specifically test or correct for environmental poisoning by chemicals or heavy metals like lead, however, so I question whether infectious disease is *the* most important factor in IQ determination. Nonetheless, their study shows it to be a potent factor.

Further buttressing the importance of infection's effect on IQ, Eppig points out that it is also a powerful predictor of average state IQ within the United States. High-IQ states include Massachusetts, New Hampshire, and Vermont, while California, Louisiana, and Mississippi bring up the rear. The five states with the lowest average IQ share higher levels of infectious disease than the five states with the highest average IQ, and this relationship holds across all of the states in between.[40]

The next year, Christopher Hassall and Thomas Sherratt of Carleton University repeated the New Mexico study using a more sophisticated statistical analysis. They confirmed that "infectious disease may be the only really important predictor of average national IQ."[41]

The parasite-stress theory of intelligence also helps explain why, as mentioned in Chapter 1, nations like Kenya have seen large IQ gains accompany improvements in health status. Just as adding iodine to the diets of 1920s Americans raised the IQs of "dullard" areas by 15 points, the average Kenyan IQ rose 14 points between 1984 and 1998 in parts of the country that enjoyed improvements in health, including reduced disability from untreated infectious disease. Parasite stress explains this, but genetics does not: the IQ gains were too high and too rapid to be explained by evolutionary and genetic mechanisms.

This infection connection also helps to explain similar patterns of IQ elevation, like the "Flynn effect," in developed nations like the United States (see Chapter 1). These wealthy nations are largely free of the rampant infectious disease found in the developing world, except where pockets of neglected tropical diseases (NTDs) and other infections plague poorer communities of color.[42] In the United States, areas that *are* plagued by such infection are also areas where people of color live and areas of depressed IQ ratings. Like the Kenyan increase, the Flynn effect in the West can't be explained genetically.

Neglected tropical diseases most frequently undermine health and cognitive vigor in areas of the United States where conditions resemble those in the developing world. African American children in Baltimore share a life expectancy with those born in Nepal. As the *New England Journal of Medicine* reported in 1990, black men in Harlem are less likely to live to the age of sixty-five than men in Bangladesh, one of the world's poorest nations.[43] The report added that "similar pockets of high mortality have been described in other U.S. cities." Such disparities persist.

National averages denoting a welcome decline in mortality rates distract us from the fact that they don't decline for everyone. Vast differences exist between neighborhoods, and hence between races, within the same area and even within the same city. In 2007, when I lived on 106th Street in a predominately Hispanic area of East Harlem, it had one of the lowest life expectancies in NYC. But ten blocks away, on Ninety-Sixth Street on the Upper East Side, the life expectancy was—and continues to be—the highest in the city. Ten years later, this discrepancy persisted. Fully 650,000 New Yorkers live in communities where the death rate for African Americans under sixty-five is twice the rate for white Americans. Their health mirrors the health of people in the developing world, where infections unknown to wealthy Westerners run rampant. Hotez illustrates one reason why: the environments are similar.

> Why, I can take you to areas such as the historical African American wards of Houston such as the fifth ward, or other areas and show you conditions of extreme poverty that closely resemble Recife (in Brazil)—dilapidated housing, no window screens, discarded tires filled with water and organic debris—it looks like the global-health movie we might show to first-year medical students, but it's Houston.[44]

As with lead, air pollution, and other toxic exposures, the distribution of infectious disease in America is inextricably linked to the tangled factors of poverty and race. Such intelligence-eroding pathogens do preferentially affect the poor. But within the ranks of the poor, it is racial-minority groups who sicken most often and fare the worst. Texas, home to the nation's three poorest metropolitan areas and Ground Zero for diseases like Zika and Chagas, has 4.5 million people living below the poverty line, the largest number of any U.S. state. The poverty rates are highest among Hispanics (26 percent) and African Americans

(23 percent). Poor Americans are at risk for NTDs, but poor Americans of color are at the highest risk.

Emerging Infections on the Brain

Infectious diseases haunt racial-minority populations. HIV/AIDS, which damages the brains of fetuses and adults alike, is probably the best-known infection that disproportionately affects black and Hispanic communities. We've long known that HIV is a disease of poor people of color here, just as it is in Africa. Seventy-one percent of Americans with HIV disease are black or Latino, and 53 percent of the people who died from HIV/AIDS in 2013 were African American. HIV dementia is a well-known example of the damage the virus wreaks when HIV crosses the blood-brain barrier to contribute to various types of neuronal injury.

Less well known are other direct psychological and cognitive costs of HIV infection, which often attacks the brain. The extent of the cognitive damage it causes in children is even more profound. HIV crosses the blood-brain barrier to injure neurons, so HIV-positive children risk a spectrum of brain dysfunction, from encephalopathy to developmental delays.

Nearly nine of every ten U.S. children living with HIV are African American or Hispanic. Infected children in the United States suffer from lower-than-average memory, speed of processing, and verbal comprehension.

Infants who acquire HIV prenatally from their mothers risk developmental brain dysfunction, from encephalopathy to subtle cognitive impairment, language disorders, and developmental delays.[45]

A 2010 longitudinal study of more than three hundred such HIV-positive children by University of Southern California psychiatrists found that they fell into the low-average scale for memory, speed of processing, and verbal comprehension.[46]

According to the researchers, neurodevelopmental problems in children and adolescents with HIV might be linked to "changes it provokes in proinflammatory monocytes" of the immune system.

HIV is just one of many brain-damaging neglected infections linked to poverty.[47] "12 million Americans suffer from at least one NTD, and many of those 12 million are impoverished African Americans," Hotez told a House Energy and Commerce committee in 2016.[48]

Between 2003 and 2005, the poorest areas of Houston were the hardest hit by a mysterious outbreak. Doctors at the Texas Children's Hospital and the National School of Tropical Medicine discovered that the culprit was the mosquito-borne dengue fever, a virus that had been all but eradicated from the United States by spraying with DDT, a probable human carcinogen, in the 1950s. At the time of the outbreak, dengue fever was regarded as a "foreign" NTD. However, the United States banned DDT in 1972, paving the way for dengue fever's eventual resurgence here.[49]

When it reappeared, not one Houston patient was accurately diagnosed with dengue, including the two people who died.[50]

"Poor whites," too, have often been referred to as an ethnic group—Appalachians, for example, as discussed in Chapter 1—when it comes to discussion of their inferior intelligence, and they too sometimes suffer disproportionate infection. They also suffer disproportionately from mind-crippling intestinal worms, such as the threadworms (including *Strongyloides stercoralis* and *Ascaris lumbricoides*), that cause impaired childhood development around the globe, just as African Americans in some regions suffer from worm infestation.[51]

African Americans suffer far higher death rates than whites and die an average of four years younger than white Americans do. But poor marginalized whites are slowly losing their lead. In a *Proceedings of the National Academy of Sciences* report, David

Squires and David Blumenthal write that mortality rates in the United States for white Americans aged twenty-two to fifty-six rose between 1999 and 2014. For thirty years before that, death rates had fallen by 2 percent a year.[52] What's more, Squires and Blumenthal cite suicides, drug overdoses, and alcohol-related liver disease as the drivers of this excess mortality, calling them "despair deaths."

Books that address these rising death rates (still much lower than that of African Americans) are beginning to appear, but I've found almost no contemporary literature that explores relationships between their supposedly lower intelligence and environmental exposure, as I have found for people of color. I suspect there is a link, but the data vacuum in this area prevents me from discussing it as I had hoped.

The prevalence of pathogens and infectious diseases found among pockets of racial minorities contributes to these disease vectors' invisibility in the United States. So does a comforting public health narrative claiming that the United States is protected by disease surveillance and environmental measures. We have confidence in the ability of the wealthy United States, with its rich matrix of health care and clean, climate-controlled environments, to ward off NTDs in a manner that the tropical world cannot.

Periodic trash removal, environmental regulations, and prevalent air-conditioning are common here and discourage the breeding conditions for mosquitoes and other disease vectors.

However, such measures are often insufficient or absent in neighborhoods of color. Laws that maintain safe environments are inadequate, unenforced, or simply don't exist.

This is especially apparent in the nation's poorest major metropolitan area, McAllen-Edinburg-Mission, Texas, which is situated near America's eight poorest smaller cities, all of which are marred by poor sanitation, environmental contaminants, and overcrowded substandard housing. These are home to ethnic enclaves of African Americans and other minority-group members. They are also located in the warmest part of the nation, which adds to the risk of infection for the poor people of color who live there.

We see the same institutional failures around the country, including in African American enclaves of Houston, where Robert Bullard, the author of *Dumping in Dixie,* has documented how sanitary services and even basic utilities are often missing from African American communities, and not just the poor ones. Even middle-class African American neighborhoods sometimes lack basic utilities and adequate sanitary services.

We're not talking just about rural backwaters but also of thronged urban sites, which is significant because many infectious illnesses thrive and gain virulence in the sort of overcrowded conditions so commonly found in cities.[53] This environmental neglect of poor communities of color has laid out the welcome mat for infectious diseases. Dilapidated housing infested with triatomine bugs that carry Chagas disease, rodents carrying hantaviruses, cockroaches that have been shown to worsen asthma, desultory garbage collection and disposal, water and microbes that collect in run-down air conditioners, to say nothing of old tires that provide breeding grounds for mosquitoes, foment infectious disease in the United States, just as they do in Haiti, South African townships, and Brazil.[54] In this sense, the oft-invoked distinction between diseases that threaten the developing world and those of the developed world has eroded.

Chagas disease is borne by an unwelcome immigrant—the triatomine or "kissing" bug, which lives in the cracks of substandard housing and passes on the parasite to people by defecating while sucking their blood. When the victim scratches the affected area he transfers the pathogen-laden fecal matter into the tiny bite wound triggering a chronic, silent parasitic infection that can lead to fatal heart or intestinal damage in two of every five sufferers. It also causes intellectual retardation in as many as one in ten sufferers.[55] Chagas mainly affects Hispanic communities.[56]

Cysticercosis

Cysticercosis causes epileptic seizures and other brain damage in a process as gruesome as any horror film.

Beginning in 2008, lurid national headlines screamed, "The Worms That Invade Your Brain," "Worm Removed from Woman's Brain," and "Hidden Epidemic: Tapeworms Living Inside People's Brains." MRIs of patients who presented to emergency rooms with sudden epilepsy or fainting began to reveal that their brains were irregularly studded with tapeworms.

That's right, tapeworms. Although we think of them, when we must, as infesting the human digestive system, where well-nourished specimens can grow to twenty feet or more, these tapeworms result from a parasitic infection known as taeniasis.

Each tapeworm produces a wealth of 50,000 eggs, which are shed in the feces of infected people. Once on the ground and eaten by pigs, they grow into larvae that normally burrow into porcine blood vessels where they wait to be consumed by humans eating undercooked pork, and their life cycle begins again.

But sometimes the eggs from the body of an infected person take a fateful detour when accidentally ingested by another human instead of a pig. When the infected person prepares food without washing his hands, for example, an egg develops into a larva that burrows into the human bloodstream and hitches a ride to the brain. Tunneling into the brain, these larvae become encysted, cloaking themselves from the immune system with specialized tissues. Thus ensconced and unmolested by the immune system, they unleash the horribly versatile disease called cysticercosis.

Cysts near the brain's visual cortex can blind the carrier. Cysts near the language area can disrupt speech or its comprehension. Cysts sometimes block the flow of cerebral fluid, causing hydrocephalus, which necessitates a shunt to relieve the pressure and prevent unconsciousness. Blindness, epilepsy, and lowered mental function (which means lowered IQ) are common, and so is death.

Treatment may not save the intellect because although the drug praziquantel kills the larvae, it also unleashes a vigorous immune response—"friendly fire"—that ends up harming the brain.

Cases are more common than one might think. In 2012 Ted Nash, chief of the gastrointestinal parasites section at the National Institutes of Health, told *Discover* magazine, "Minimally, there are 5 million cases of epilepsy [worldwide] from neurocysticercosis." Between 1,500 and 2,000 neurocysticercosis cases are diagnosed in the United States[57] every year when confused, unconscious, or epileptic patients are brought to the hospital and the detection of antibodies definitively identifies the disease.[58]

Most cases are found in Hispanic Americans, who are more likely to ingest pork-borne tapeworm eggs. A 2012 *Public Library of Science (PLOS)* report reads, "It is now well established that cysticercosis is a leading cause of epilepsy among Hispanics living in Texas, with Texas and California most likely representing the greatest share of the 169,000 cases of cysticercosis in the U.S."[59]

As with many such worm-borne pathogens, deworming arrests the cognitive decline, but it does not restore lost intellectual functioning. Today, one of every ten people brought to Los Angeles hospitals with an epileptic seizure suffers from neurocysticercosis. This is only one dramatic manifestation of an epidemiological sea change: tropical diseases—and their neglect—are not limited to the tropics anymore.[60]

Toxocariasis

Toxocariasis is caused by *Toxocara canis,* a parasitic roundworm that infects dogs. People can acquire it from soil and sandboxes contaminated with dog feces.[61] The larval worms navigate through the lungs and brains of children to cause pulmonary dysfunction and wheezing akin to symptoms caused by asthma, but they also cause cognitive and intellectual deficits.[62] Toxocariasis slipped over the border to infest poor, run-down urban areas and crumbling rural homes in the American South.[63] *Toxocara canis* is now carried by 21 percent (more than one in five) of African Americans, for a total of 2.8 million people.[64] "The fact that it may affect the mental health of so many black children has prompted me to speculate that toxocariasis might be responsible for educational achievement gaps during preschool and the school-aged years," Hotez said in 2013.

That year, his speculation was validated by the heavily detailed annual U.S. National Health and Nutrition Examination Survey (NHANES), which used the Wechsler Intelligence Scale and the Wide Range Achievement Test to compare the mental acumen of children aged six to sixteen who were infected with *Toxocara canis* with that of uninfected children. It found that children without the infection scored considerably higher on both intelligence-test scales, even after correcting for a laundry list of potentially confounding factors, like "socioeconomic status, ethnicity, gender, rural residence, cytomegalovirus infection, and blood levels of lead, all of which have been implicated in affecting intelligence test scores."[65]

African Americans and Hispanics also have the nation's highest rates of death from *Toxoplasma gondii* infection, which causes the disease toxoplasmosis and a slew of other medical problems.[66] This parasite of cats can result in fetal death and abortion as well as psychiatric syndromes that include neurocognitive deficits and even schizophrenia, according to Robert

Yolken, director of developmental neurovirology at Johns Hopkins University.[67]

Trichomoniasis

An estimated 3.7 million people in the United States suffer from *Trichomoniasis vaginalis,* a parasitic infection that causes the nation's most common curable STD. "Trich" is a largely silent infection; fewer than one in three infected people notice symptoms. But it can cause fetal death and damage, including neuronal damage that cripples intelligence, and its rate is ten times higher among black women than others. Twenty-nine percent of African American women carry T. vaginalis, not too far from the 38 percent of infected women in Nigeria. This means that black women are ten times as likely as white or Hispanic women to harbor the parasite, which increases the heterosexual spread of HIV. Highly sensitive diagnostic tests can now detect T. vaginalis, and it can be easily cured with a single dose of metronidazole. Unfortunately, neither the test nor the treatment is routinely administered.[68]

Fortunately, most people infected with cytomegalovirus (CMV) never know it, because it rarely causes serious symptoms or problems in healthy people with functional immune systems. But a pregnant woman who develops an active CMV infection can pass the virus to her baby, and one in five infected babies suffer impaired nervous system development, leading to hearing and vision loss that may become severe and permanent, as well as mental disabilities. If the infection is diagnosed at birth, medications called antivirals can be given, which can alleviate the visual and hearing loss.[69]

CMV infects more African American women than mothers of any other race.[70] Infected adults often have no symptoms, except for those with compromised immune systems who can experience symptoms and even die.[71] A myriad of other infections like malaria, cerebral tuberculosis (which targets the brain), hookworm, trachoma, and leishmaniasis preferentially impair the brains of ethnic minority groups in the United States.[72]

Climate of Fear

But why have NTDs gained a foothold in the United States? Xenophobes may accuse immigrants of bringing these transplanted nightmares north with them, but this is inaccurate: many NTDs are infectious, but not contagious. This means that they are transmitted to others by pathogens, but not by infected people.

Instead, blame the U.S. climate,[73] because many microbes function within a narrow temperature range, and parasite life cycles often require heat. "The U.S. is somewhat unusual in being a wealthy nation much of whose population lives in very warm, humid regions," Stan Cox, a senior scientist at the Land Institute, told the *Washington Post* in July 2015.[74] U.S. temperatures are warmer than those in Europe and most of the affluent West.

Accordingly, scientists predict that global warming will hasten the spread of pathogens and disease. As the climate grows even warmer, microbes and disease vectors, such as the snails that carry schistosomiasis, the sandflies that carry leishmaniasis, and triatomine bugs, will expand their territory, and so will the *Aedes aegypti* mosquitoes that disseminate Zika, dengue fever, chikungunya, and yellow fever.[75] These are already common on the Gulf Coast, as is the domestic Zika vector *Aedes albopictus,* which is found in the eastern United States.[76]

Meteorological events also escalate the risks of infectious diseases that threaten intelligence. In 2005, the "kissing bugs" that carry Chagas disease and the snail populations that cause schistosomiasis proliferated in the aftermath of Hurricane

Neglected Parasitic Infection	Selected Prevalence Data	Major Risk Factors	Clinical Sequelae in Adults and Children 1	Clinical Sequelae in Adults and Children 2	Congenital Clinical Sequelae	References
Toxocariasis	>21% Seroprevalence among African Americans (up to 2.8 million African Americans); >17% seroprevalence in the American South	African American race, male sex, poverty, low education level, lead ingestion, contact with dogs, coinfection with *Toxoplasma*	Neurologic and psychiatric: cognitive delays; epilepsy; ocular manifestations	Pulmonary: diminished lung function, asthma	Not well established	[13–22]
Cysticercocis	Up to 41,000–169,000 infected persons; likely widely underrecognized, with only an estimated 1,000 hospitalized cases diagnosed	Hispanic immigrants	Neurologic: epilepsy; chronic headaches	None	Not well established	[24–28]
Chagas Disease	300,167 cases	Hispanic Americans	Cardiovascular: cardiomyopathy neurysms; conduction disturbances; sudden death	Gastrointestinal: megaviscera	Congenital: congenital Chagas disease syndrome	[29–38]
Toxoplasmosis	1.1 million new cases annually, including 21,505 cases of ocular toxoplasmosis and up to 4,000 cases of congenital toxoplasmosis	African American ethnicity and poverty	Neurologic and psychiatric: cerebritis; schizophrenia; bipolar and other mood disorders	Ocular: retinitis and retinal scars and other ocular findings	Congenital toxoplasmosis syndrome: hydrocephalus; chorioretinitis; intracranial calcifications; cognitive deficits; hearing loss	[39–47]
Trichomoniasis	7.4 million new cases annually	African American ethnicity — 10 times more common	Genitourinary: vaginitis; pelvic inflammatory disease; pregnancy complications	HIV coinfections	Neonatal infections	[48–52]
Total	Approximately 12 million incident or prevalent infections, including some people living with chronic sequelae					

Features that produce clinical manifestations and sequelae similar to selected noncommunicable diseases.
doi:10.1371/journal.pntd.0003012.t002

Katrina,[77] and we can expect similar spikes after each hurricane season.

But humans cause problems after weather emergencies, too. Three Caribbean-U.S. hurricanes, two severe Mexican earthquakes, and waves of flooding across Bangladesh, Nepal, and India made the autumn of 2017 meteorologically memorable. Rebuilding will take years, and unfortunately post-disaster construction has a way of disproportionately worsening environmental conditions for the marginalized people in affected areas,[78] as debris is dumped in their neighborhoods.

Rebuilding often uses toxic materials as well. In post-Katrina New Orleans, for example, environmental and air-pollution standards were relaxed to accelerate reconstruction. What's more, as part of post-Katrina recovery efforts, private companies were allowed to acquire public housing. The homes of poor evacuees were condemned as nuisances, marked for demolition, and resold at extremely cheap prices. Such actions, along with the billions allocated to the Army Corps of Engineers for rehabilitation of levees, entrenched, rather than eased, the vulnerability of poor communities of color, both by introducing additional environmental exposures in the short term and by displacing former residents from the safer, rehabilitated housing in the long term.

What happened in New Orleans is reminiscent of San Francisco's attempt to move Chinatown from the city center to a more peripheral area on the city's outskirts, ostensibly as part of its rebuilding efforts after the 1906 earthquake.

A 2017 *Nature* article by Benjamin K. Sovacool points out that Hurricane Harvey hit hardest in poor areas and minority communities located near the Arkema chemical plant in Crosby, Texas, which exploded after the storm.[79]

Even in the absence of floods and hurricanes, U.S. Geological Survey scientists warn that the armies of pathogens on the ground enjoy ample air support as "hazardous bacteria and fungi hitchhike across the Atlantic on [15,000-foot-high] winds, eventually scattering the pathogens of the developing world over American yards and playgrounds."[80] (You can see this for yourself by searching "North African dust plumes" at https://earthobservatory.nasa.gov.)[81]

New Domestic Threats

Not all emerging diseases in the United States hail from Africa or the global South. Some are homegrown.

In October 2014 a *Proceedings of the National Academy of Sciences* report describes how Johns Hopkins professor of medicine Robert Yolken was surprised to find a waterborne virus called *Acanthocystis turfacea* chlorella virus—mercifully nicknamed ATCV-1—lurking in the throats of two of every five of his Baltimore research subjects.[82]

The cognitive tests conducted as part of his study delivered an even greater shock. When the performance of the infected was compared to that of those who did not harbor the virus, researchers found that the infected made calculations 10 percent more slowly and displayed shorter attention spans, suggesting that the virus may retard the ability to calculate and to process visual information.[83] This reduced mental functioning occurred independent of potentially confounding factors like age, socioeconomic status, education, place of birth, and smoking status. Gender and race made no difference. No demographic data allow us to stratify this threat by race, but Baltimore is 65 percent African American.

Yolken's findings persisted when this small study was repeated in a larger population. This correlation strongly suggests an intellect-lowering role for ATCV-1 in humans, but a potentially definitive study was not conducted: the scientists did not expose healthy people to the virus in order to compare their performance pre- and post-infection for obvious ethical reasons.

However, the team did test a group of mice before and after exposing them to ATCV-1, and found 1,000 gene changes in brain regions that are integral to memory and learning. The infected mice also wore dunce caps, taking 10 percent longer to navigate a maze than uninfected controls, and they spent 20 percent less time investigating novel environments, which suggests a reduced attention span. Critics suggested that the researchers had found not an IQ-lowering microbe but rather sample contamination, but Yolken refuted this suggestion in an article in the February 2015 *Proceedings of the National Academy of Sciences.*[84]

Red Tide

Other pathogens endemic to North America are known to diminish mental function. When I worked in a poison control center during the 1980s, we kept track of warnings of "red tide" algae that periodically bloom during warm weather. We used the information to warn worried callers and emergency-room physicians of this occasional threat and also urged newspapers to advise their readers. Shellfish toxins sicken 60,000 every year, and kill nine hundred,[85] but only after I looked up a new disorder associated with the algae—"amnesiac shellfish poisoning"—did I understand how cruel and unusual red-tide poisoning could be.

In 1987, 107 Canadians fell victim to amnesiac shellfish poisoning after consuming tainted mussels from the waters off Prince Edward Island (PEI).

Eating seafood tainted by *Pseudo-nitzschia* algae triggers much more than the usual mayhem of nausea, intestinal pain, projectile vomiting, and explosive diarrhea. For the PEI diners, these unpleasant effects were eclipsed by horrifying cognitive symptoms. Thanks to domoic acid, a potent neurotoxin contained in *Pseudo-nitzschia* algae, the infected became confused, aggressive, disoriented, and prone to endless crying jags. They also permanently lost the ability to form any new memories.[86]

This toxin is a neural imposter that resembles the essential amino acid glutamate so closely that our brains cannot discern the difference. When taken up through the blood-brain barrier, it kills neurons in the hippocampus, the seat of memory, and in the amygdala, which mediates fear and anxiety, generating prolonged crying that is sometimes followed by coma and death. Four of the PEI victims died, and in the spring of 1991, domoic acid was also measured in razor clams collected on Washington State beaches. *Pseudo-nitzschia* has been identified in seven algal species and has spread to contaminate shellfish in Japan, Denmark, Spain, Scotland, Korea, and New Zealand.[87] Other toxic algae, like *Pfiesteria piscicida,* also induce mental deficiencies, such as memory loss.

Pseudo-nitzschia red tides have recurred with increasing frequency, and scientists predict that climate change will trigger even more frequent occurrences. No data suggest that people of color are more likely than others to come into contact with these brain-eroding algae. Instead these appear to be a small and localized, but growing source of potential intellectual deterioration for everyone, and another argument for the EPA to acknowledge and counter climate change.

Mad Cow and More

And then there are prions, infectious proteins first identified by Stanley Prusiner, who won the Nobel Prize in 1997 for his discovery. These cause bovine spongiform encephalopathy (BSE), better known as mad cow disease, and its analogous human diseases. The latter include kuru, discovered decades ago among the New Guinea Highlanders, and Creutzfeldt-Jakob disease

(CJD),[88] a rapidly progressive, fatal disorder marked by profound mental deterioration. Its symptoms include impaired memory, loss of mental acuity, dementia, and impaired muscle control.

CJD killed legendary choreographer George Balanchine—after being initially misdiagnosed as Alzheimer's disease when he could no longer remember dance moves and musical scores or maintain his balance long enough to demonstrate new choreography. According to the NIH, CJD strikes only three hundred U.S. residents each year, but Laura Manuelidis, chief of neuropathology at Yale Medical School, has theorized that many cases of supposed Alzheimer's, which can be definitively diagnosed only upon autopsy, are misdiagnosed CJD, just as Balanchine's was. Because African Americans are diagnosed with Alzheimer's twice as often as whites, scientific scrutiny into this possibility could be a great boon to the community because at least some cases of CJD, unlike Alzheimer's, can be prevented by removing suspect meats and proteins from the diet.

Psuedo-nitzschia, the ATCV-1 chlorovirus, and prion disease are obscure hazards, but another domestic malady, Lyme disease, is familiar, and so are the cognitive deficits some infected people suffer.

A lesser-known tick-borne disease, Powassan virus (POW), causes fever, vomiting, seizures, and memory loss, and about half of its survivors are left with permanent neurological symptoms. According to the Centers for Disease Control and Prevention, 10 percent of all cases are fatal.[89] Symptoms usually show up about a week to a month after the tick bite. No vaccines or medications are available to treat or prevent the virus infection, so the best way to protect yourself is to avoid contact with wooded areas that may harbor ticks, and to apply DEET repellents. A 2018 *Consumer Reports* survey found that only one in three Americans thinks that DEET is safe, but since 1960, 90 percent of reported problems were mild and only one in every million users has experienced a seizure or serious medical problem. Most of the latter occurred when people misused the product, failing to follow the package directions. The EPA does not classify it as a carcinogen, nor have CDC studies of adults and children found health hazards.

Still, experts advise not using DEET around children younger than two months of age, and pregnant women should avoid it out of an abundance of caution, although no study has demonstrated any effects of small exposures on unborn children.[90]

We must remember that exposure to microbes doesn't happen in a vacuum. The communities that are most susceptible to infection are those that also suffer onslaught by poisonous heavy metals, toxic chemicals, and other agents whose devastating effects on fetal and adult brains have been demonstrated. The effects are additive and perhaps even synergistic, rendering existing calculations about the relative contribution of microbial infection to intelligence simplistic.

The collective freight of these exposures conspires to drag down the intelligence of ethnic minorities.

Mind-Boosting Microbes?

One potential answer to the destruction wrought by infection is still speculative, to be confirmed by future researchers—if it is ever confirmed at all. Although it remains under investigation, it is too intriguing to ignore.

We know that some microbes erode intelligence, so perhaps there are microbes that enhance it. A pair of immunologists think that they have found one.

In the 1990s, John Stanford and Graham Rook patrolled the shores of Lake Kyoga in Uganda seeking the answer to a puzzle: local residents had responded much better than others to the BCG tuberculosis vaccine, and they wanted to know why. They found the answer in the soil of the lake bed: *Mycobacteriam vaccae,* a bacterium with immune-modulating qualities that make the vaccine more potent.[91]

M. vaccae has proved versatile. Injecting it into mice raised their serotonin levels and decreased their anxiety. The investigators wondered whether it might affect learning, so they fed the bacteria to mice and then tested how well they navigated a maze. The bacteria-fed mice raced through the maze twice as quickly as the controls.

Another scientist tested *M. vaccae* in humans in an attempt to prolong the lives of lung cancer patients: it did not prolong their lives, but it tamed their anxiety and raised their spirits. Recently, *Environmental Health and Technology* published a study of twelve hundred people showing that *M. vaccae* might significantly enhance learning ability.[92]

Vinpocetine is another brain-boosting candidate, a synthetic compound that is derived from vincamine, an alkaloid found in the *Vinca minor L.* plant.

It has been used clinically in many countries for more than thirty years to treat cerebrovascular disorders such as stroke and dementia[93] by clinicians who think it improves brain function.

Dr. Akindele Olubunmi Ogunrin, a neurology researcher at the University of Benin Teaching Hospital in Benin City, Nigeria, studied the efficacy of vinpocetine (trademark name Cognitol) in improving memory and concentration in cognitively impaired patients. He found that vinpocetine was in fact effective in improving memory and concentration in patients with epilepsy and dementia, although its efficacy was minimal in demented patients.[94]

According to WebMD, "Vinpocetine might have a small effect on the decline of thinking skills due to various causes," and no significant harmful effects were reported in a study of people with Alzheimer's disease who were treated with large doses of vinpocetine (60 mg per day) for one year.

Vinpocetine is sold by prescription in Germany under the

brand name Cavinton, and it is already on sale as a dietary supplement in the United States under various names.[95] But I'd advise you to save your money. Although website advertisements claim that "more than a hundred" safety and effectiveness studies have been funded by the Hungarian manufacturer Gedeon Richter, few double-blind controlled clinical studies have been published. And of these, "Most...were published prior to 1990, and results are hard to interpret because they used a variety of terms and criteria for cognitive decline and dementia." So for now, the prospect of using a microbe to enhance IQ instead of diminish it remains an intriguing but unanswered question.

What, then, can be done to prevent the damage inflicted by microbes or to mitigate the effects of poisonous air, water, metals, food, and chemicals? In the next chapter I suggest solutions for you, your family, and your community that offer hope for maximizing the intellect of our nation.

PART III

Mission Possible: How to Bolster the Nation's IQ

CHAPTER 6

Taking the Cure: What Can You Do, Now?

Start where you are. Use what you have. Do what you can.

— Arthur Ashe

"Ever since I caught the lead, I've been messed up in the head. I can't control my anger or feelings," David confided to a reporter.* "I could have been better than I am."[1]

David, who is now twenty-one, was severely poisoned as a child in his Milwaukee home. He suffered chronic poisoning over several years, although as little as a single chip of lead paint might have landed him in the hospital where doctors were able to save his life, but not all his intellect.

The outlook is bleak for some lead-poisoned children, but there is hope for many. I was pleasantly surprised to learn from lawyers like Evan K. Thalenberg and Thomas Yost of Baltimore that the cognitive skills and school performance of some poisoned children they represented have improved dramatically. Some have even gone on to attend college after winning settlements that provided them funds enough to afford counseling and educational support.

* David is not his real name.

However, most people profoundly affected by environmental poisoning do not have such resources. This makes preventing poisoning extremely important. Children—and adults—need protection from lead, mercury, arsenic, industrial chemicals, pesticides, air pollution, and even from environmentally related diseases that can impair the brain.

The only known national solution is to eradicate harmful, underregulated poisons from residential housing, schools, water, food, and fence-line communities, and many scientists and activists have been working for decades to achieve just this.

A rollback of large-scale U.S. poisoning requires more than their knowledge and dedication, however. It also requires political solutions. A healthy environment—breathable air, potable water, food and game that are not imbued with heavy metals, homes that are not permeated with intellect-robbing industrial poisons, soil without deadly pesticides—is not something individuals and communities can create without the force of law and government support.

To be sure, protecting the brains of exposed Americans means banishing, not reducing, the sea of dangerous pollution in which they have been forced to live, study, and work. Ending pollution means forcing powerful industries to act against their financial interests and this cannot be accomplished by individuals. It is the responsibility of our government, including the EPA and the public health professionals that advise them, to eradicate untested, underregulated poisons from residential housing, schools, and fence-line industries.

The requisite legislation and regulation can be enacted only by a strong, active EPA that is dedicated to protecting Americans, not industry. We need to update regulations to account for the latest scientific findings about, for example, the special vulnerability of young brains to minute exposures. And we need to enforce them. Only this will create a safe environment for all, not only for the wealthy and powerful.

This critical work cannot wait for more research that demonstrates the harms that pollution is wreaking in communities of color. More than enough evidence has been amassed for us to act. Our nation must embrace the precautionary principle so that protection of the citizenry takes precedence over amassing "enough" research to satisfy the polluters and their scientists.

If future research exonerates a restricted or banned chemical, it always can be restored to use; but it is much harder—if not impossible—to restore the millions of IQ points lost to the effects of chemicals presumed to be safe.

Poisoned communities can be restored to health, too. Despite a long history of industrial malfeasance and governmental apathy (and worse), there is hope. Solutions exist for the problem of environmental poisoning, and some communities of color have found them by uniting.

We would do well to realize that this is a marathon, not a sprint. The greatest chance of success depends upon enlisting the support of researchers who have shown themselves committed to the intellectual health of communities of color and on forging partnerships with other environmental groups. Although mainstream environmental groups have long focused on preserving nature, conservation, and recreational issues, their acknowledgment that the planet needs protectors, not exploiters, means that they and poisoned communities of color have much in common. Working together they have already achieved some successes.

Yet, while we await environmental sanity, there are steps that individuals can take to fight for a less toxic environment and higher intelligence, and I discuss many of these here.

Enroll Your Child in Pre-K

Pre-K enrichment programs like Head Start have been demonstrated to improve the academic performance of children. Other forms of prekindergarten or pre-K enrichment also help children practice and improve verbal and reasoning skills that are invaluable in the classroom and that boost IQ and standardized test scores. Free pre-K programs should be mandatory nationwide as one way to level the academic playing field from the beginning. Until they are available everywhere, find one for the children in your life at the Head Start website, where you can check your child's eligibility and apply (https://www.benefits.gov/benefits/benefit-details/1928).

Fight Toxins in Schools

Unfortunately, many children face their highest risks of environmental poisoning in what should be the safest of all venues: at school. A 2001 study by the Center for Health, Environment and Justice entitled "Kids at Risk—Toxic Schools: Creating Safe Learning Zones" revealed that more than 600,000 mostly poor and minority-group students in Massachusetts, New York, New Jersey, Michigan, and California were attending nearly twelve hundred public schools located within half a mile of a federal Superfund- or state-identified contaminated site.[2] Contact the EPA if you are concerned that your child's school may be one of them.

In large cities like New York and small communities like Anniston, schools as well as homes have been found to have harbored noxious, brain-damaging chemicals and heavy metals like lead. Because children spend so much time in schools, such environments require attention and immediate detoxification.

Unfortunately, you cannot always rely upon the school administrators to assume leadership. In New York City, school administrations hid their schools' lead contamination by deliberately misleading investigators and workers who were sent to test school lead levels.[3] If you ask for information about contamination issues in your child's school and do not receive answers, be prepared to persist and to go higher.

The EPA offers information about attaining a lead-free school at https://www.epa.gov/dwreginfo/lead-drinking-water-schools-and-childcare-facilities.

There is power in numbers, so try to recruit other parents in your search for answers and solutions. Start at the EPA. Its guide to training testing and its "Guide for Community Partners," a blueprint for organizing, can be downloaded from https://nepis.epa.gov/Exe/ZyPDF.cgi?Dockey=P1004WLH.txt. (See also Chapter 7 for a complete list of practical steps to help you organize for the environmental health of your community.)

Poison-Proof Your Home

You cannot eliminate all sources of toxic exposure, but there are steps you can take to reduce and minimize exposures to environmental poisons within your home.

Air quality. If you live in an area plagued by heavy industrial emissions, the air quality in your home is not completely under your control, but you can improve it. If you can afford to, keep your doors and windows closed and use your air conditioner to minimize emissions, at least during high-traffic hours. Some energy companies and cities have home-energy rebate programs to assist with bills.

Vermin control. Cockroaches and dust mites worsen asthma; rodents carry pathogens that may encourage hypertension, which is linked to lowered cognition. To remove them, hire a professional exterminator and follow his advice to keep these unhealthy visitors at bay. If you rent, check your municipal codes or with the local housing authority or Legal Aid to determine your rights as a renter. Usually it is the landlord's responsibility to ensure vermin-free housing, so she must pay for extermination.

Vacuuming. Use a HEPA vacuum often to minimize your family's exposure to tracked-in toxic substances, dust mites, cockroach parts, and other vermin.

Cleaning supplies. Many cleaners contain volatile toxic chemicals such as halogenated hydrocarbons that harm the lungs and present neurotoxic threats that harm the brain. Others, such as bleach and ammonia, or bleach and various acids, become poisonous when mixed together. Read labels, and whenever possible, choose cleaners that do not contain a long list of chemicals—many are hydrocarbons or poisons that are readily absorbed through the skin.

Diluted bleach (which should always be handled with care), simple detergents, and ammonia can be used, separately, for many household cleaning jobs and are far more economical than complex specialized cleaners. But always use gloves and handle bleach carefully, because it can cause a lot of damage if it's splashed on the body or if it gets in the mouth or eyes.

All cleaning products should be stored in locked, child-safe cabinets: a bad taste will not deter toddlers from sampling these poisons.

Many people use essential oils in cleaning, but these are sometimes toxic as well. Some should not be used by pregnant women, and they can be expensive. For a useful description of low-toxicity cleaning options, see *Less Toxic Living: How to Reduce Your Everyday Exposure to Toxic Chemicals—An Introduction for Families,* by Kirsten McCulloch.[4]

Paper masks. Use paper masks during periods of highest exposure to poor air quality. This might include outdoor activities near toxics-spewing factories, riding the subway in an area of substandard air quality, or outdoor tasks that may heighten your exposure to polluted soil, air, or other pollution "hot spots." Recall the story from Chapter 3 of Shirley Baker, who donned a paper mask to mow her toxics-soaked lawn. City dwellers should emulate commuters in heavily polluted Asian cities who are frequently seen wearing masks on public transportation. Studies show they indeed offer some protection against not only pollutants but also some communicable diseases like the flu and colds.

Shoes. Lead, industrial chemicals, animal dander, pesticides, chemical dust, pathogens, and a wide assortment of uninvited visitors can hitch a ride into your home on the soles of your shoes. Consider leaving your shoes at the door and going shoeless inside, or trading your shoes for flip-flops when you enter. And ask your guests to do the same. But avoid walking barefoot outside, where you can absorb pollutants through your skin and pick up parasites, including hookworms, which have been demonstrated to sap intelligence and are making a resurgence in parts of the United States.

Poison-Proof Your Water

Between polluted bodies of water, corroded external and indoor pipes, and even the overuse of corrective chemicals like chlorine, few of us can be certain that the water we drink in our homes is free of dangerous contaminants. We need only remember that the people of Flint were repeatedly assured that their water was safe while their brains were being assaulted by high levels of lead to realize that we should be vigilant.

There are many options to protect yourself from waterborne contaminants, but none is foolproof and very few counteract every kind of toxic exposure.

Bottled water. Bottled water is relatively safe, although it is an expensive and inconvenient solution, especially for a family. The expense incurred goes beyond the financial outlay. The energy required and pollution generated by its processing, bottling, transportation, and disposal makes bottled water an environmentally questionable option.

Furthermore, the plastic bottles that contain the water may

not be perfectly safe, but there is a way to check. You may have noticed numbers and symbols on plastic containers; these can tell you whether the bottle that holds your water contains chemicals that are known to be or suspected of being toxic.

Look for triangular recycling signs on the bottle. If the triangle contains the number 1 and the letters PET or PETE (for polyethylene terephthalate), it means the plastic does not contain BPA, which is good. But PET is a form of phthalate, which should be avoided whenever possible. Do not re-use such bottles because the chemicals can eventually leach into water and food.

Bottles with the number 2 inside the triangular recycling symbol and the letters HDPE (high-density polyethylene) are a better choice because they contain no BPA, phthalates, or any other known toxic chemical.[5]

Water filters. There are many types of water filters, from $20 pitcher filters to $1,500 systems that are permanently installed in your home plumbing. The various models use everything from activated charcoal to ultraviolet (UV) light to remove pathogens, parasites, heavy metals, and chemicals. But few, if any, filters remove all of these or remove them completely.

If your household includes infants or pregnant women, seriously consider using a water filter. As noted in Chapter 4, young children drink about four times as much as adults relative to their weight, and the young, especially fetuses, are much more sensitive to toxic pollutants than adults.

What kind of water filter should you invest in? That depends on the contaminants you wish to remove, your budget, and how much space you have.

First find out what issues your water is known to have, bearing in mind that both the water that enters your home and your household pipes may be sources of contamination.

You can do this by checking the Environmental Protection Agency's water-quality reports for your city at epa.gov/safewater. These are updated every July. But if you use well water or if your house was built before 1986 when lead-free plumbing was mandated, you should have your water tested.[6] Call your local health department to ask for a free test kit (the EPA website on the previous page lists local laboratories), or call the EPA Safe Drinking Water hotline at 800-426-4791. If the lead levels are below 150 ppb, a water filter can remove them. If they are higher, ask the EPA or your local health department for guidance.

Once you know what contaminants threaten your water, decide whether you want a point-of-use (POU) filter, which can be used in a pitcher or installed on the spigot, or a point-of-entry (POE) filter attached to your home plumbing where the water enters your home.

Bear in mind that installing a POE system will filter the water entering your home but will not remove pollutants emanating from your own plumbing, so unless your home was built after 1986, lead or other toxic substances can leach from your pipes into your water. Also, even if you do not have lead plumbing, lead can enter your water from other sources, such as the solder on pipes.

If you rent, you probably want a portable system because POE systems are expensive.

No matter which type you choose, change filters at least as often as the manufacturers' directions indicate in order to keep them effective and to prevent the growth of bacteria and mold.

Following are the most common types of filters:

- **Carbon filters** are filled with carbon-based material like charcoal or burnt bamboo or coconut. These materials are "activated" to increase their ability to adsorb contaminants. They remove chlorine, pesticides, and petrochemicals, but do not remove all lead and fluorine unless they incorporate a filter impregnated with "KDF filters," granules that are composed of a half-zinc, half-copper alloy.
- **KDF filters** include zinc-copper granules that remove heavy metals, ions, bacteria, algae, and fungi. They do

not remove pesticides and parasites. If they get clogged, they can release pollutants back into your water. So they need to be backwashed with hot water. These filters are most useful when paired with the carbon filter.

- **Ion-exchange filters** use a variety of resins to remove heavy metals, nitrates, and fluoride. They do not remove sediment, pesticides, microbes, or chlorine. Bacteria may grow on them, which seems to defeat one purpose of a water filter.
- **Reverse osmosis filters** are complicated, expensive, built-in filtration systems that remove pesticides, fluoride, petrochemicals, chlorine, asbestos, nitrates, heavy metals, and radium. Some people avoid them because they worry that the resulting water is slightly acidic, but this is the normal pH of most natural water sources and is not dangerous to health.
- **Sediment filters** are made from a wide variety of materials that remove sand, rust, clay, and dirt particles. They are often used with carbon filters in order to keep them from clogging too quickly and often.
- **UV sterilizers** are the same type used in home aquariums. They are not filters, but instead use ultraviolet light to kill pathogens and algae. However, most treated water already contains chlorine, which also kills these pathogens. This makes UV systems redundant for most people.

Once you have decided upon the type of water filter you want, look at independent quality ratings. Not only will they tell you which filters performed well in tests, but these ratings also give valuable information about product recalls and other issues.

For example, in the late 1990s, when I worked as an editor at a national consumer magazine, I discovered that some water filters actually leached lead into the water they "purified"! Save yourself such unpleasant surprises by checking sites like *Consumer Reports* (www.consumerreports.org/products/water-filter/ratings-overview), *Good Housekeeping* (www.goodhousekeeping.com/health-products/g684/water-filters), or Reviews.com (www.reviews.com/water-filter).

Fluorine. Many filters remove fluorine, which raises the question of whether you *want* the fluorine removed from your water. Fluorine is used to prevent tooth decay, the incidence of which has continued to fall since it was first added to U.S. water. But approximately fifty studies worldwide, including several observational studies, have found an association between small fluorine exposures and lower-than-average IQ. This has caused some researchers to fear that fluorine can harm the developing brain in fetuses and the very young, but the question is hotly debated. Recent large studies in China and Mexico have found an association between lower IQ and prenatal fluoride exposures. In Mexico, pregnant women who had higher levels of fluoride in their urine, and presumably delivered to their fetuses, gave birth to children who had lower IQ scores when tested between four and twelve years of age than women with lower levels of urinary fluoride. As the researchers themselves point out, such observational studies can demonstrate a possible association but not cause and effect. Most U.S. researchers I consulted, like pediatric dentist Donald Chi, professor of oral health sciences at the University of Washington, see no poisoning risk from the levels in U.S. water sources.[7] More precise studies, perhaps longitudinal ones, must be done to better characterize fluoride's possible cognitive harms and provide parents more useful answers.[8]

A movement to ban water fluoridation in the wake of the new studies cited above is under way over the objections of the EPA and American Dental Association. In April 2017, the anti-fluoridation group Fluoridation Action Network along with allied medical and dental groups filed a lawsuit in California seeking to ban water fluoridation. The EPA responded by asking the federal court to dismiss the suit, but in December 2017 a federal judge refused to do so.[9]

Even if fluoride's effect on IQ is supported by well-conducted future studies, a general ban may not be nuanced enough. Depending on what future studies reveal, perhaps a fluoride ban should focus more narrowly on fetuses and newborns. It is possible the small amounts of fluorine found in fluoridated American water are harmful to the brains of fetuses and the very young but perfectly safe for older children or adults. And because the fetus does not have teeth that fluoride can protect, there may be no advantage to this earliest fluoride exposure. Health experts will need more research data before they can tell us the logical steps to take.

This recommendation may seem inconsistent with my recommendation that we embrace the precautionary principle by erring on the side of banning industrial chemicals until their safety has been demonstrated. However, fluoride is a special case: unlike the noxious industrial chemicals in question, fluoride conveys demonstrated health benefits—except in fetuses and neonates who have no teeth to protect.

Meanwhile, you still must make a decision for yourself and your family, bearing in mind that children of color have more dental issues on average than other children (partly because Medicaid policies limit access to quality dental care).

You should seek advice on this question from your child's doctor and dentist. If you decide to avoid fluoride altogether, make sure you buy a water filter that will remove it, and give your young child training toothpaste, which does not contain fluoride.

If you decide that you want fluoride's dental benefits but worry that your filter may have removed it, ask your doctor about using naturally fluoridated salt such as Himalayan salt or the fluoridated salt sold in parts of Europe. Bear in mind that dentists recommend that children use only a pea-size amount of toothpaste; more is unhealthy and increases their exposure to fluoride, possibly causing dental fluorosis—discoloration of the tooth enamel.[10]

Food and Nutrition: Provide a Poison-Free Start

As described in Chapter 4, getting safe, adequate nutrition while avoiding brain-draining contaminants like mercury is especially challenging for pregnant women, but it can be done.

Multivitamins. Women who may become pregnant can give their babies the best, poison-free start by enriching their dietary habits now, before they conceive. Start by conferring with your doctor, because the advice in this chapter is general and does not apply to any one individual's medical needs and condition—only a doctor can provide specific advice for you. If your doctor agrees that a daily multivitamin will provide the nutrients you and your baby need, this is what you should take, although you should also ask whether to take additional folic acid to prevent neural-tube problems and choline supplements to prevent alcohol from damaging your baby's brain. You should also consider adding fish or other sources of omega-3 oil to your diet. If your doctor agrees, two 6-ounce servings of (low-mercury) fish weekly (see "Seafood Safety" below) will boost brain health for you and your unborn baby.

Breastfeeding. Consider breastfeeding, which, as explained in Chapter 4, conveys many benefits for a child's brain as well as body. But if you live in a fence-line community or other toxics-ridden area, be sure to ask your doctor specifically about whether you should change your diet or whether it is safe to breastfeed at all, because some poisons can taint breast milk. In addition to your pediatrician, organizations like La Leche League International can advise you about breastfeeding. Breastfeeding confers so many advantages to a baby's physical health and mental acuity that it is sometimes advisable even if a poison can enter breast milk. In Chapter 4, I explained that the future IQ of breastfed babies was higher than normal despite the effects of the mercury in their breast milk. In 1993, breastfeeding researchers followed the cognitive development of breastfed children and reported in *Early Human Development* that even though they received more contaminants through breast milk than bottle-fed children, the breastfed scored significantly higher on the Bayley and McCarthy mental and psychomotor development tests (at all time points from two years through five years) and had higher English grades on report cards from grade three or higher.[11]

Alcohol. To protect any future offspring from alcohol-caused brain damage, it is safest to abstain from alcohol completely if you could become pregnant, especially in light of the rate of underdiagnosed fetal alcohol brain damage in African American, First Nations, and Hispanic infants. Recent studies have questioned the purported health benefits of a daily serving of alcohol, and many experts now think that abstaining may be healthier than moderate consumption.[12] But if you choose to drink, no more than a single daily serving is best—and *daily* serving is the key: decades of studies show that this is not the same thing as seven drinks on Saturday night. If there is a possibility you could become pregnant, ask your doctor whether she advises avoiding alcohol completely or whether limiting yourself to a single drink daily is sufficient.

Baby food. As Chapter 4 revealed, many baby food products, even organic ones, test positive for lead and arsenic, including 80 percent of infant formulas. And that's not the only brain-damaging contaminate they may contain.[13] Consider making your own baby food. If you buy commercial baby food, research it first to see what brands and types are safest. The Clean Label Project, which is supported by grants, donations, and its certification program, is a good resource.

Fruits and vegetables. Wash and peel fruits and vegetables to minimize pesticide residue that can bioaccumulate and cause cognition problems in children.

Processed foods. Worried that the chemicals listed on a label could compromise your children's brain development? Choose minimally processed foods to avoid adulterants that could affect your family's health, and avoid overprocessed food. Using minimally processed foods also allows you to control the ingredients of the food you prepare.

Preservatives. Avoid most preservatives. This is trickier than it sounds, because not every preservative is harmful: some preservatives are necessary to protect your health by keeping microbes at bay. For example, buying peanut butter without any preservatives can be risky because harmful molds called aflatoxins that can cause liver cancer and delayed development in children can grow in peanut butter. But preservative chemicals can cause health problems, so the wisest course is to use as many unprocessed foods as possible and therefore avoid the need for many preservatives. It is also good to understand what the common preservatives are and whether they can pose hazards.

To do this, you don't have to become a chemist, but you do have to know where to look for information. Your doctor may be able to help or to recommend a nutritionist or dietician who can. The habit of reading food labels is a good one to adopt, because all preservatives added to food must be declared on the label's ingredients list.

The government labels many food preservatives GRAS, which means "generally recognized as safe." It reasons that these chemicals have been used and monitored for long periods without serious concerns arising. But GRAS also means that they have not been specifically tested for human safety, so it is impossible to be sure they are safe, and especially whether they are safe for mental development, which the chemicals are not typically monitored for.

Here is a list of common GRAS preservatives and their purposes:

antimicrobial agents—these kill bacteria and stop the growth of mold on foods

antioxidants—these prevent the oxidation or spoilage of food and offer health benefits by reducing harmful free radicals in the body

benzoates—the salts of benzoic acid

butylated hydroxyanisole (BHA)—a waxy solid used to preserve butter, lard, meats, and other foods (linked to some cancers)

butylated hydroxytoluene (BHT)—similar in structure and function to BHA, but in powder form (linked to some cancers)

chelating agents—these prevent spoilage by bonding with the metal ions in certain foods

citric acid—found naturally in citrus fruits

disodium ethylenediaminetetraacetic acid (EDTA)—used in food processing to bind manganese, cobalt, iron, or copper ions

nitrites—the salts of nitrous acid

polyphosphates—used as anti-browning agents in dips and washes for peeled fruits and vegetables[14]

propionates—the salts of propionic acid

sorbates—sorbic acid and its three mineral salts: potassium sorbate, calcium sorbate, and sodium sorbate

sulfites—a group of compounds containing charged molecules of sulfur compounded with oxygen, including sodium sulfite, sodium bisulfite, sodium metabisulfite, potassium bisulfite, and potassium metabisulfite

vitamin C (ascorbic acid)—a water-soluble vitamin and its salts, sodium ascorbate, calcium ascorbate, and potassium ascorbate
vitamin E (tocopherol)—a fat-soluble vitamin

Although they are on the GRAS list, questions have been raised about the safety of BHA and BHT, which have been linked to cancer. Sodium nitrite and nitrate, often used to preserve meats like bacon and sausage, are strongly suspected of causing cancer and should be avoided. So should cadmium: although it is on the GRAS list, it is also a toxic chemical.

Home canning. Canning, also called "preserving" or "putting up" food, is an economical way to preserve large quantities of garden fare and meats for storage by packing them into glass jars. When done correctly, this allows you to avoid both spoilage and artificial preservatives. Canning got many of our forebears through harsh winters and other times of scarcity.

But canning must be performed meticulously in order to avoid spoilage and fatal food poisoning. It is very dangerous if not performed correctly. While working in a poison center, I learned that most cases of the fatal food poisoning botulism arose from eating not restaurant or supermarket foods but improperly processed home-canned fare.

Normal cooking temperatures do not kill botulism or denature its toxin, so canned foods that harbor it are deadly. A mere drop of tainted food can kill. Botulism is not self-limited like many other types of food poisoning, and it can kill quickly. Even if a person is rushed to the hospital and given the "antidote"—botulism antitoxin—they cannot always be saved, especially if too much time has elapsed, and the antitoxin cannot undo damage that has already been done.

Because botulism is fatal, don't take chances. Don't trust the canning instructions in recipes and cookbooks, because they are not always sufficient to prevent botulism. Unless you are willing to commit to purchasing the right equipment and following the U.S. Department of Agriculture (USDA) instructions[15] *exactly*, find another way to save money and avoid preservatives in foods.

You can order the USDA Complete Guide to Home Canning, by calling 1-888-398-4636 or download it from http://nchfp.uga.edu/publications/publications_usda.html.

For more advice, contact your state or county extension service, which you can locate by clicking a link at https://www.cdc.gov/features/homecanning/index.html.

If you are unsure whether a food, either home-canned or commercially purchased, is safe, don't take chances: Discard it. Better to lose a few dollars than a life. And don't taste-test it! It takes only a drop of botulism-tainted food to kill.

If you have already eaten food and have concerns about its safety, call your local poison center. You should tape the number to your home's telephone or refrigerator, especially if you have children. If you don't know it, call the national poison control center, which can be reached at 1-800-222-1222: it is always open, seven days a week, twenty-four hours a day.

Seafood safety. Seafood is a common source of food poisoning, partly because it must be carefully prepared and stored to prevent microbial growth that can send a victim to bed or the bathroom for hours of unpleasant symptoms. But worse, as I discuss in Chapter 5, a few types of seafood poisoning seriously harm the brain, causing permanent short-term memory loss, confusion, and even death.

If you fish, avoid such brain-threatening illness by paying close attention to "red tide" warnings in the newspapers, online, or by calling your local extension service, which you can find at http://pickyourown.org/countyextensionagentoffices.htm. You can also go to the National Ocean Service at https://oceanservice.noaa.gov/news/ and search for "red tide."

To avoid mercury poisoning, limit yourself to two or three servings of fish a week. Choose smaller fish of these types: light tuna (avoid white albacore tuna, which has a higher mercury content), salmon, tilapia, pollock, catfish, shrimp, and mackerel (pick smaller mackerel fish, which will have lower concentrations of mercury). Avoid shark, swordfish, king mackerel, and tilefish, which the FDA lists as having the highest levels of mercury. Avoid eating whale and dolphin at all costs; their mercury levels are extraordinarily high.[16]

Dollar-store fare. Buying food in dollar stores is an easy way to save money—unless your purchase is tainted by toxins, heavy metals, or other poisons that harm the brain. Be especially wary of imported candies and tinned meats. Lead and microbes can leach into them from their containers, and these dangers won't be acknowledged on the label. Imported pottery, too, can have high levels of lead: such purchases might be better obtained from discount outlets that carry U.S.-regulated goods. Join with friends and neighbors to buy economically in bulk at a big-box store instead, where you are less likely to encounter unregulated foreign fare.

Iodized salt. Because iodine deficiency remains the greatest global cause of mental retardation, buy only iodized salt, and avoid sea salt unless it is infused with iodine, because processing removes its natural iodine. If you adopt a low-salt diet for health reasons, ask your doctor to recommend alternative sources of iodine. Some European salt and Himalayan salt contains fluoride, additional amounts of which should be avoided by pregnant women and very small children. Water in the United States is already fluoridated, and more than fifty studies suggest that additional fluoride may cause an IQ drop in the very young.

Supplements. Both the World Health Organization (WHO) and the United States Dietary Supplements Health and Education Act (DSHEA) of 1994 define dietary supplements as products other than tobacco that supplement the diet, such as vitamins, minerals, herbs, botanical products, and amino acids.

In addition to iodine, some nutritional supplements and vitamins are known to improve cognition. Some can also improve the cognition of unborn children when taken by their mothers. Folic acid prevents neural malformations like spina bifida; vitamin D prevents bone disorders and a slew of other conditions, and it is especially important for dark-skinned people, who are more likely than others to suffer a vitamin D deficiency. Evidence suggests that other supplements are worth investing in: choline may help to prevent both fetal alcohol disorders and Alzheimer's disease, for example.

If you are one of the 16 percent of Americans who use both alternative therapies and conventional medicine,[17] you must involve your doctor. In 1993, David Eisenberg published a study in the *New England Journal of Medicine* (that was later confirmed by his subsequent studies) revealing that most—three of every four people—who use alternative medicine hide this fact from their doctors.[18] For the sake of your and your family's health, don't be one of them.

Some herbs and supplements that don't harm adults can devastate an unborn child's brain. Moreover, if your doctor doesn't know what you are taking, her prescriptions and recommendations may be working at cross purposes with the alternative agents, robbing them of their potency. Or your alternative treatments and your doctors' prescriptions may potentiate each other, dangerously heightening the effect of your medications. If your doctor absolutely rejects the use of alternative medicine and refuses to advise you about it, consider finding a doctor who is more alternative-friendly.

It is true that we would be better off if we could obtain all our needed vitamins from food rather than from pills, but eating enough of the right foods to obtain all these daily nutrients is not easy: it would probably be a full-time job, and quite expensive. The right supplements provide convenient nutritional insurance.

However, be careful in choosing nutritional supplements. They are very loosely regulated, tested only *after* they are suspected of harming someone, and their advertised claims do not always match up with the clinical reality. Some supplements contain lead or even prescription medications that are not listed on the ingredients label. Also, the people who distribute and sell them can be dangerously naïve about pharmacology, which has caused some serious injuries when the wrong configuration of a nutritional chemical was sold. The clerk in the health food store who persuades you that a supplement will help alleviate your child's lead-poisoning symptoms cannot help you if it sends your child to the hospital instead.

Here are just a few of the most common supplements that have caused problems after being sold with medical claims. Many are sold under several names and this list is far from exhaustive, so the fact that a product does not appear here does not mean it is safe.

Aconite is traditionally known as wolf's bane, and is derived from a genus of plants that have long been used to address inflammation, joint pain, or gout. But aconite can also harbor poisonous chemicals. In fact, it is responsible for most cases of severe herbal poisoning in Hong Kong.[19]

Caffeine powder addresses weight loss, increased energy, and athletic performance but it is dangerously potent: one teaspoon equals 28 cups of coffee, according to the FDA, which has banned some brands after two users died from caffeine overdose in 2014. (A safe dose for most adults is 200 mg, or one-sixteenth of a teaspoon.)[20]

Chaparral is made from a Californian shrub and addresses weight loss, inflammation, colds, rashes, and infections but is harmful to the liver and can cause hepatitis.[21]

Coltsfoot is an herb that addresses coughs, sore throats, laryngitis, and asthma. But it contains alkaloids that cause tumors and was banned by the German government after an infant died because her mother had consumed coltsfoot tea during pregnancy.[22] The ban was amended to allow the sale of a variety developed without any alkaloids.

Comfrey is an herb used to alleviate heavy periods, stomach problems, and chest pain, but it was banned for internal use in the United States because it causes liver damage and is unsafe for use by pregnant women.[23]

Germander is used for weight loss, fever, arthritis, gout, heart failure, and stomach problems, but it also causes liver damage when used often. More than 45 people were sickened by it in France, and one died.[24]

Greater celadine, a member of the poppy family, is used to treat stomachaches, pain, and cancer, but it is toxic even at moderate doses and can cause fainting, convulsions, and paralysis.[25]

Green tea extract powder addresses weight loss but it can also dangerously elevate heart rate and blood pressure and injure the liver.

Kava reduces anxiety but also raises the risks of Parkinson's disease and paradoxically, of depression.

Lobelia addresses respiratory problems and is used as a smoking-cessation aid but the risks of nausea, diarrhea, rapid heartbeat, seizures, coma and, possibly, death make it too dangerous to use.

Mä huang, also known as ephedra or "Mormon tea," has caused seizures, psychosis, myocardial infarction, cardiac arrhythmia, stroke, and death when given in high doses or for long periods. It has been used as a stimulant to enhance athletic performance, but the FDA banned it from supplements in 2004 after several athletes' deaths were linked to it.[26]

Methylsynephrine is a stimulant favored by those who want to increase energy, lose weight, and gain greater athletic stamina. But these come with the risks of heart-rhythm abnormalities that can lead to cardiac arrest.

Pennyroyal oil alleviates breathing problems and aids digestion but is also tied to nerve damage, convulsions, and liver damage.

Red yeast rice is used to help lower high cholesterol and prevent heart disease but it is tied to hair loss and it can dangerously increase the effect of cholesterol-lowering statin drugs.

Yohimbe is taken for everything from depression and weight control to erectile dysfunction. But it raises blood pressure, causes seizures and liver damage problems, and may cause panic attacks.[27]

"Natural" is not a synonym for "safe." Alternative supplements, teas, and other nostrums are often embraced by people who believe that they are natural and therefore safer than a pill or medicine that has been artificially derived in a laboratory and manufactured in a factory.

But this is an illusion.

You don't have to spend much time in a poison center or even in the wild to realize that remedies found in nature, like Nature herself, can be extremely powerful, and even deadly. Herbs and supplements can sicken and kill, especially when used or prepared inappropriately, in excessive dosages, or for a prolonged period of time. But even experts have fallen victim after underestimating the potency of natural remedies. In 2015, the *British Medical Journal* recounted how a "trained herbalist" in the UK collapsed and was brought to the hospital after she dosed herself with an herb to combat her insomnia. Doctors identified it as the potent poison belladonna, which is derived from a plant colloquially called deadly nightshade.[28] In the early 1980s, a Russian mushroom expert died while visiting a Syracuse, New York, college after consuming a poisonous mushroom that closely resembled a benign edible species in the USSR. The herb *Stephania tetrandra* was used in Belgium in the early 1990s for weight loss. But in some cases another herb, *Aristolochia fangchi,* was mistakenly substituted, causing at least one hundred cases of renal failure.

Unlike pharmaceuticals, herbal and "natural" dosages are not standardized. And variations in soil quality, growing season, method of preparation, and even geographic location all cause large differences in a given medicinal plant's potency, making it hard to know how much to take.

A survey of ginseng products found that some products were two hundred times more potent than others that were similarly named. A similar survey of an ephedra product found that some batches contained ten times more ingredients than others produced by the same company.

Furthermore, an analysis of 2,609 herbal samples found that 23.7 percent—almost one in four—were adulterated with pharmaceuticals. A parallel California study found that 7 percent contained undeclared pharmaceuticals.[29]

Avoid Heavy Metals

I've discussed the dangers of small doses of methylmercury, such as those derived from fish in the diet, that can have devastating effects on the nervous system and brain, lowering the IQs of many exposed people and injuring fetuses. But heavy metals like mercury cause concern elsewhere, too.

Thimerosol/mercury in vaccines. A form of mercury called thimerosol has been used as a preservative in some vaccines given to both children and adults. Although the *Lancet* study by Andrew Wakefield that claimed to find a connection between the preservative and autism was recalled (with its author discredited and struck off the British medical roll), many fear that thimerosol or even the vaccines themselves may cause autism in their children.

A thorough discussion of this issue would require far too much space to engage in here, but thimerosol has been removed from all but a very small handful of vaccines and is not currently used in any that are given to children. It poses no autism risk.

Mercury in batteries. Another common source of mercury is in the small, round, flat batteries, many the size of a penny or smaller, in devices like watches and hearing aids. Children and toddlers sometimes eat them, but so do adults. Some people accidentally ingest a watch battery while holding it in their mouth while changing the batteries: Don't do this.

Older adults, especially those suffering from eyesight problems or dementia may mistake them for pills. Unfortunately, the ingestion sometimes is not discovered until symptoms set in and the victim begins vomiting, drooling, retching, and suffering from abdominal pain, rashes, dark stools, irritability, and fever.

This is a medical emergency: send the person to the hospital immediately. Do not "wait and see." Bring the device that originally contained the battery so hospital staff can determine exactly what type of battery was consumed. Do not give the victim ipecac, salt solution, or anything else to induce vomiting: this can worsen the damage, possibly forcing a battery that has entered the stomach to become lodged in the esophagus where it can impede breathing.

There is also a battery ingestion hotline at 202-625-3333, which you can call if you need immediate help, or if circumstances prevent you from taking the person to get professional medical help. But do not call the hotline instead of taking the person to the hospital immediately. At the hospital, staff may wait and watch the progress of the battery through the system; they may retrieve it with the aid of an endoscope; or they may administer medication such as dimercaptosuccinic acid (DMSA) or dimercaptopropane sulfonate (DMPS) n-acetyl penicillamine.

To prevent such exposures, securely tape the battery compartments on children's toys and electronics shut. Store batteries in childproof containers if you have young children in the house, and lock them away if you have elderly people living with you—they can so easily mistake the batteries for pills.

Mercury amalgams. Many people worry that dental fillings made of an amalgam that can contain as much as 50 percent mercury may poison them or their children. Scientific reports on their safety vary. However, such studies have shown that the mercury in fillings may cross the placental barrier to harm an unborn child. Unfortunately, removing these fillings carries risks as well, exposing both the patient and the dentist to mercury vapor.

In fact, some experts think that the dental workers are at higher risk from such fillings than patients. Research on the question continues, so ask your dentist for guidance and possible answers.[30]

Workers besides those in the dental industry risk mercury exposure, too, especially those employed in coal-burning industries and in gold mining.[31] In ancient Rome, criminals sentenced to work in quicksilver mines had a short life expectancy because of the toxicity of the mercury in the cinnabar they mined.

Some laboratory workers today risk exposure to the very dangerous compound dimethylmercury, which is so toxic that a drop spilled on the skin, even on a gloved hand, has caused death.[32]

Some governments have heavily regulated the use of mercury. The European Union, which has historically exported but not imported large amounts of mercury, has prohibited the export of mercury since March 2011. One stated reason is that although poisoning is uncommon in the EU, it is a problem in native Arctic communities, China, India, and in many developing countries.[33]

Lead. If you and your family are exposed to lead via tainted water, food, peeling indoor lead-based paint and dust, soil, or industrial emissions in your neighborhood, any effective, permanent solution must come from regulatory changes imposed upon the responsible industry or municipal government. But you are not powerless. You can take steps on your own, and banding together with neighbors and advocates will offer some further protection and hasten governmental responsibility.

To protect your child, first find out what year your housing (or your child's daycare or school) was built. If it was built after 1978, that's good news, because federal law has prohibited the use of lead-based interior paint in homes since that year.

The Centers for Disease Control and Prevention also have recommendations:

Children don't spend all their time at home so be sure to tell babysitters, grandparents, and other caregivers not to allow your children near peeling paint, window jambs, or "chewable surfaces" painted with lead-based paint.

If your home was built before 1978, have it tested for lead. If the test is positive, move if at all possible. If you cannot move, consider boarding young children with a friend or relative while you find out how to remedy the situation. Contact your health department about your options and ask a lawyer how your landlord can be compelled to abate the lead.

In some cities like New York, tenants can notify their landlord directly if they see chipping paint. They can submit a complaint to the city's 311 system, a public number that lets people report non-emergency issues. The city Department of Housing Preservation and Development (HPD) will respond. To see what similar laws may apply in your community, contact your municipal housing department or the state EPA. To find information about laws, resources, and how to make complaints almost anywhere else in the nation (except tribal reservations), go to https://www.epa.gov/lead. For tribal matters, contact Environmental Protection in Indian Country at https://www.epa.gov/tribal.

Some cities provide an online database of lead-imbued housing that prospective renters can check. In the case of Baltimore, the Maryland Department of the Environment database at http://mde.maryland.gov/pages/index.aspx informs you whether a listed property is free of hazards like peeling, flaking lead paint and lead dust. Its website also details the state laws pertaining to lead poisoning, and a section that explains how to file a complaint. New York state maintains a similar site at https://www.health.ny.gov/environmental/lead/programs_plans/index.htm.

Caregivers should regularly wash children's hands and toys and monitor children as closely as possible to minimize their putting things in their mouths that may carry lead dust. They should also wet-mop floors and windowsills frequently, but these are only partial, temporary measures. Nothing fully protects your home's occupants except full abatement of lead by a professional.

Children and pregnant women living in pre-1978 homes should not be present during renovation or lead-abatement activity that disturbs old paint. In New York City and some other areas, lead-based paint was banned before 1978, so older paint may be safe in certain places. If you have any doubts have it tested.

Home lead tests are not your best option. At about ten to twenty-five dollars each they are economical, but their complexity makes many brands difficult to use accurately. In some models, colors indicate the result, so that some do not work if you are color-blind.[34]

Instead consult your health department to conduct a test

for lead in your home. Or find an EPA-certified lead testing professional—do not allow anyone without this certification to do the work—at epa.gov/lead/pubs/renovation.htm. If you own your home, government-insured loans may be available to help pay for this service that can cost hundreds of dollars.

Pesticides. A pesticide is any substance or mixture of substances intended for preventing, destroying, repelling, or mitigating any pest. All pesticides must be tested, registered, and carry a label approved by the EPA. Despite the agricultural community's regular use of pesticides, homeowners are the number one users. Pests take many forms besides insects.

Categories of pesticides include:

algicides—including some pool chemicals
fungicides, miticides, larvicides, and more
germicides—bathroom disinfectants
herbicides—plant defoliants and desiccants
insecticides—insect attractants and repellents, flea collars for pets
mildewcides—contained in some cleaning products
rodenticides—rat and mouse killers

Understand the label warnings. "Caution" identifies the pesticides that are slightly toxic—the least harmful. "Warning" tells you it is more poisonous than a pesticide with a "Caution" label, but still carries moderate toxicity. Don't assume that products lacking the skull-and-crossbones logo are nonpoisonous. "Danger" or "poison" on the label indicates that the pesticide is very poisonous, highly irritating, or toxic. Handle and store them carefully, following the label directions concerning safety equipment and protective gear. Call your poison center for instructions if anyone is exposed by oral, inhalation, or skin contact. Do not give ipecac or anything else to induce vomiting unless instructed to do so by your doctor or a poison center employee.

Make Safe Household Purchases

From children's toys and furniture covered in illegal lead paint to water filters that leach lead to recalled household items that still appear on store shelves, checking the SaferProducts.gov website is a fast, reliable way to protect your family from hazardous or recalled items with a simple search. The site, managed by the federal Consumer Product Safety Commission, also allows you to report hazardous items.

Even after following all this advice, your power to mitigate the harmful effects of environmental poisons is limited. Real safety for you, your family, and your community lies in ending, not moderating, toxic exposure, and this is a goal that requires joint effort. The next chapter shows how to organize your community against polluters.

CHAPTER 7

A Wonderful Thing to Save: How Communities Can Unite to Preserve Brainpower

Never doubt that a small group of thoughtful, committed citizens can change the world; indeed, it's the only thing that ever has.

— Margaret Mead

Even without significant wealth or political clout, communities of color acting jointly have begun to transform the environment and health of marginalized people assailed by invisible poisoning. Consider that the first sweeping demands for environmental justice were ignited by groups like the middle-class residents of nondescript Afton, in North Carolina.

On a hot day in mid-August, throngs of North Carolinians from this predominately African American town and surrounding Warren County hoisted signs and banners as they chanted and sang stanzas of "Ain't no stopping us now," "We shall not be moved," and other familiar hymns of the civil rights movement. White supporters brandishing signs and fists swelled their ranks, and formations of chanting neighborhood children filled the roads, hoisting placards and banners.

The marchers had gathered to defy a state convoy of trucks.

As the protesters vowed to halt the vehicles' incursion, armed helmeted police responded by aiming weapons as the sheriff took up his microphone to invoke law and order.

He warned that if they did not disperse, the assembly would be arrested. As the first resister was dragged, passively limp, into a police van, others lay in front of the trucks. As they were cuffed and led away, one by one, the chants and songs continued into the night.

The placards, the shouted demands, the playlist of venerated civil-rights hymns, the armed police, and the arrests presented a drama familiar to Americans. Throughout the 1960s, we absorbed such spectacles nightly with our dinners on the six o'clock news.

Not only the human-rights passion play but also the demographics were those of an iconic Southern civil rights protest, in

Police drag a passive Afton, North Carolina, protester to jail. (AP Photo/Steve Helber)

which the black and disenfranchised ascended a moral high ground in the face of state force.

Seven of every ten Warren County citizens are black, and one of every five is poor, but although most residents were middle-class, they wielded little political power. However, thanks to their collective action, Afton's protests assumed a grander scale than most. National television cameras transmitted a seemingly endless stream of protesters and arrests that continued for *six weeks.* During that time, 550 people were arrested, giving rise to what the *Duke Chronicle* called the largest Southern civil disobedience since Martin Luther King Jr. had marched through Alabama.

Birth of a Movement

But Dr. King was long dead. The Afton protests took place not in 1965, but in 1982, and the participants were not resisting voting disenfranchisement, unequal education, or racial segregation. They were fighting the trucking of PCBs (polychlorinated biphenyls) into a landfill sited in their neighborhood.

Since the 1920s, PCBs have been valuable in industry because they are resistant to acids, bases, and heat, which makes them useful as insulation in electrical equipment such as transformers (devices that use induction to transfer electrical energy between circuits) and capacitators (electrical components that store potential energy in an electric field) as well as lubricants. Although their toxic effects led to their U.S. manufacture being banned in 1979, PCBs from other countries and those that persist in the environment still pose a threat to Americans. They linger in many other products and applications, from flame retardants to paints, coatings, adhesives, and inks.[1]

Marchers' shouts demanded and their placards read "No PCB," "Landfills Kill," and "We Care About Our Future: Don't Harm Generations to Come."

After the victories of the civil rights movement, including the end of de jure racial segregation, African American residents of neglected communities took aim at a different kind of racial sequestering. They began to demand equal access to clean land, food, air, and water through lawsuits, political demands, and civil disobedience. Afton was one of the first communities of color to fight its selection as a dumping ground for garbage and toxic wastes. It certainly was the most influential: it is no exaggeration to say that the people of Afton ignited the global environmental justice movement through their civil disobedience. It was an idea whose time had come.

As the state made good on its threat to jail the protesters, the sheriff's invocation of law and order was ironic: it was Afton's residents who were the victims of interstate criminal activity.

The crime spree had begun in the middle of a night in late July when Robert Burns of Jamestown, New York, drove his specially outfitted truck to North Carolina under cover of darkness.

Civil rights activist Benjamin Chavis leads protesters against illegal PCB dumping in the mostly African American town of Afton, North Carolina. (AP Photo/Greg Gibson)

Accompanied by his two grown sons, he made the illicit pilgrimage to save money by surreptitiously spewing 31,000 gallons of toxic oil along Carolinian roadsides instead of disposing of it safely and legally. These midnight forays continued for two weeks along more than two hundred miles of the state highways. Burns left meter-wide swaths of earth contaminated with carcinogenic PCBs, dioxins, and dibenzofurans.

The poisoned oil had come from the Ward Transformer Company of North Carolina, one of the nation's largest transformer repair companies. Its owner, Robert E. Ward Jr., was Burns's business partner. When the illegal dumping was discovered, Burns was sentenced to three to five years in a North Carolina prison and Ward was convicted of violating the Federal Toxic Substances Control Act and sentenced to eighteen months in prison and fined $200,000.[2]

The state's attention quickly turned to protecting its citizens from the disease-causing chemicals, notably PCBs, which were then known to cause birth defects and liver and skin disorders and were strongly suspected of causing cancer. Although PCBs' ability to erode intellectual functioning was not yet well established, today we know that PCBs are also endocrine disruptors that profoundly damage the brains and impair the intellect of fetuses and the young, as explained in Chapter 3.[3]

"We know why they picked us"

Alarmed by the dire health effects posed by the 50,000 tons of soil soaked in a witches' brew of PCBs, dioxins, and other poisons, Governor James B. Hunt Jr. militated for its disposal in predominately African American Warren County as the "best available site."[4] However, scientific criteria, such as the distance from groundwater or likelihood of leaching, did not support his administration's description of Warren County as the "best" site.

"We know why they picked us," declared the Reverend

Luther G. Brown, then pastor of Coley Springs Baptist Church. "It's because it's a poor county—poor politically, poor in health, poor in education and because it's mostly black. Nobody thought people like us would make a fuss."[5]

But Afton residents made a widely resounding clamor, and a smart one: they recruited support from neighboring sympathizers and organizations in order to do so. The Coley Springs Baptist Church, the county chapter of the National Association for the Advancement of Colored People, and twenty-six neighbors joined them in filing a lawsuit in Federal District Court to block the dumping of PCBs.

This coalition charged that by ignoring more scientifically suitable alternative sites, EPA officials and the State of North Carolina practiced racial discrimination against the citizens of Warren County.

An unsympathetic judge, W. Earl Britt, removed the last legal hurdle between Afton and the PCB dumping when he ruled against the residents, declaring from his bench, "There is not one shred of evidence that race has at any time been a motivating factor for any decision taken by any official, state, federal or local."

This is despite the fact that Title VI of the 1964 Civil Rights Act directs that "No person in the United States shall, on the ground of race, color, or national origin, be excluded from participation in, be denied the benefits of, or be subjected to discrimination under any program or activity receiving federal financial assistance."

Neither did the EPA protect Afton's residents. Instead, it paid $2.54 million to fund the digging up and trucking of the poisoned earth to Afton, in effect sponsoring the pollution of Afton and its environs. This time, the deadly chemicals would be dumped openly, in the light of day and with the blessing of the state and the EPA.

In response, Afton's residents and their supporters took to the streets to resist the dumping, mounting one of the longest and most widely publicized mass demonstrations and igniting the global environmental justice movement.[6] The *Washington*

Post described the PCB protest movement as "the marriage of environmentalism with civil rights."[7]

Congress Takes Note

Afton inspired more than activism by other targeted communities. It inspired the research that is essential for documenting the racial nature of such deadly exposures.

Congress took notice of the protests, and the very next year the U.S. General Accounting Office reported to the Committee on Energy and Commerce[8] that three of every four landfills it looked at were located in the region's poorest communities. However, the GAO specifically declined to investigate race as a factor in the siting of industrial landfills, a prescient omission that helped elide the role of race in some environmental poisonings.

But sociologist Robert Bullard already knew race to be a determining factor in communities he had studied, such as Houston. His 1979 study had discovered that all five city-owned landfills were located in black neighborhoods. Moreover, "Six out of eight city-owned incinerators were in black neighborhoods," he recalled in a 2017 interview with the Union of Concerned Scientists. "And more than 80 percent of the garbage dumped in the city from the 1930s up to 1978 was being dumped in black neighborhoods. Black people only made up a quarter of Houston's population at the time."[9]

Bullard set out to investigate and quantify the role of race in other parts of the nation. His meticulous assessments while a professor at Texas Southern University in Houston correlated census tract and other demographic information, including race, with toxic exposures. With these data, he demonstrated the determinative role of race. In 1987, a more expansive survey by the United Church of Christ Commission for Racial Justice completed an inclusive national report revealing that hazardous waste facilities in the United States are more likely to be located in predominantly minority communities such as those Bullard studied in Dallas, Texas; Alsen, Louisiana; Institute, West Virginia; and Emelle, Alabama, than in white areas.[10] This overrepresentation of environmental hazards and the concomitant increased health risks were detailed in his many books such as 1960's *Dumping in Dixie: Race, Class and Environmental Quality.*

Within a decade, follow-up studies verified that this racial disparity had worsened. Such data quantified a racial factor in the siting of toxic wastes that had heretofore gone undocumented, and these data are essential to crafting political public health solutions that acknowledge the engineered vulnerability of African Americans and other Americans of color.

It was the organized efforts and persistence of Afton activists that drove the government interest. "These were invisible problems in invisible communities until they organized themselves and started to have their own dialogue with EPA," said Vernice Miller-Travis, of the Office of Environmental Justice advisory council.

Afton's own landfill battle has not yet been won because the PCBs that were dumped in 1982 remain.[11] But its residents put into motion a global sea-change larger than the activists could have foreseen.

In its 1994 Environmental Equity Draft, the EPA described Afton's PCB protests as "the watershed event that led to the environmental equity movement of the 1980's." The people of Afton catalyzed the birth of a global resistance movement to free communities of color from the physical and, as we now recognize, the mental fetters of environmental poisoning. Afton's playbook still holds lessons for the sickened and endangered.

In 1992 while a graduate student in Boston, I met Mary Ann Nelson, a Cambridge lawyer who was an executive of the Greater Boston Sierra Club. I had never before met an African American who held a leadership role in a mainstream environmental organization; although there are now many, there then were not nearly enough. I approved of the environmental agendas of groups like the Sierra Club and the Nature Conservancy, which included protection of wild places and endangered species; preserving water and air purity; working for a "clean energy" future, curbing climate change, and maintaining pressure on politicians and corporations to ensure healthy communities.

But for a long time, I couldn't help but think of these goals as tangential to the life-and-death struggles of people of color, whose very ability to breathe, think, and live a normal healthy lifespan is directly threatened by environmental crises. Today their interests are quite congruent and their partnerships with communities of color seem natural alliances brokered by visionaries like Professor Robert Bullard and Dr. Aaron Mair, who became the first African American elected as Director of the Sierra Club in 2015.

Conventional wisdom had long fed a misperception of African Americans as divorced from the joys and health benefits of

The early research of Robert Bullard, "the father of the environmental justice movement," revealed the nature and extent of selective environmental poisoning in neighborhoods of color. (AP Photo)

Aaron Mair (AP Photo/Rex Larsen)

nature. In 1969, an R. J. Reynolds Tobacco Company advertising memo advised against showing African Americans in outdoor settings. It claimed that "... the 'Outdoors' (hunting, skiing, sailing) is not felt to be suitable, as these are considered unfamiliar to the Negro ..."[12]

Today, this widespread misconception that African Americans do not seek and enjoy communing with nature persists, fed by a number of biases. The relative absence of African Americans in some national and public parks, swimming pools, and similar sites reflects not indifference but the past violent prohibitions against their presence. Robert Bullard points out that today a dearth of mass transit makes many such refuges inaccessible to urbanites.

But as I spoke with the Sierra Club's Mary Ann Nelson about her work, I remembered how formative my daily interactions with nature had been as a child. I felt that days spent roaming the woods and fields of Croton-on-Hudson; Fort Dix, New Jersey; and Aschaffenburg, Germany, while learning firsthand about nature contributed to both a lifelong appreciation of biology and a happy childhood. My friends from the Dominican Republic and Virgin Islands also wax nostalgic as they speak of growing up surrounded by natural beauty. We're now confirmed urbanites, but we agree that something has been lost in trading unspoiled nature for the streets of New York City and Chicago.

Recent research shows us that this yearning for natural surroundings reflects more than sentiment. We now know that exposure to the natural world is essential to a child's health and development—and to the health of the adult she will become.[13]

A Natural High

A growing body of research supports the positive connection between exposure to the natural environment in childhood[14] and better health at all stages of life.[15]

Stress—physical, psychological, or some combination of these—is an inescapable part of life. Responses to stress can be productive, even necessary, but severe or unremitting stress can damage one's physical and mental health. Various species of racialized stress have been implicated in everything from hypertension to suicide.

We most often read and hear of malignant stress arising in direct response to a stressor such as a threat or perceived threat.

But this is not the only overarching theory of stress. In February 2018, a trio of researchers investigated how living within unsafe environments generates a constant stress response. Writing in the *International Journal of Environmental Research and Public Health,* they discuss stress-fueling features of such toxic environments, such as: low social status, early (including prenatal) adversity, loneliness, and "lack of a natural environment."

They also discuss biological mechanisms for these: loneliness, for example, seems to drive up stress levels by chronically decreasing heart rate variability (HRV),[16] increasing cortisol levels, and distorting immunity mechanisms.

The stress caused by such factors can be curtailed, or "turned off" only when safety is perceived. Exposure to the natural environment is one of many factors capable of turning off chronic stress by triggering a perception of safety. The University of Leiden (Netherlands) authors argue that within this "generalized unsafety theory of stress (GUTS)" the environment provides the physiological substrate for the underpinning of some things we share with most other animals: fear of the unknown. From birth we have a low tolerance for uncertainty that results in a stress response that is always turned on—until we perceive signs of safety. Indications of safety disarm the stress response and in so doing they mitigate the risks of stress-related diseases like hypertension affective disorders and even obesity.[17]

The authors theorize that the prefrontal cortex, which usually suppresses brain structures such as the amygdala[18] and neural activities that mediate the stress response, especially the amygdala, do so only when the brain has perceived safety.[19] If safety is no longer perceived, these neurological "brakes" are lifted, and the amygdala reverts to high gear, quickly restoring the body to "fight or flight" mode. Heart activity increases, and blood pressure rises, and other stress responses resume.

The prefrontal inhibition of the stress response lasts only as long as the perception of safety, which in many urban environments is rarely for long. The authors note that chronic stress may be an especially important risk for African Americans.

A well-known example of membership in a discriminated minority with serious health consequences are African Americans in the United States who have lower life expectancy than their compatriots,[20] and it is believed that discrimination significantly contributes to this difference.[21] The lifetime burden of perceived discrimination and discriminatory harassment and/or assault is associated with higher blood pressure, especially ambulatory and nighttime diastolic blood pressure[22] and higher levels of inflammatory markers. Again, since it is unlikely that these chronically enhanced physiological effects are due to actual discrimination-related stressors alone, or to continuous conscious worrying about it, it seems more likely that for many African Americans daily life is unconsciously perceived as less safe than it is for others.

Previous studies have already suggested that exposure to natural rather than urban environments reduces anxiety and feelings of sadness, and normalizes heart rate, cortisol, and blood pressure even when confounding factors such as socio-economic status, exercise levels, noise, smells, and crowding are controlled for.[23] Even viewing pictures of unspoiled nature shows some of the physiological effects.

Combating environmental poisoning is critically important, but like everyone else, African Americans, Hispanics, Asians, and Native Americans need access to untrammeled nature to optimize our mental and physical health.

Over the past few decades, these two poles of the environmental movement have come together over common concerns. Mainstream environmental groups have shown their worth as powerful allies of environmental justice activists and have embraced its tenets. As early as 1994, the Sierra Club published *Unequal Protection: Environmental Justice and Communities of Color* and followed this in 2005 with *The Quest for Environmental Justice: Human Rights and the Politics of Pollution.* In 2013 the Sierra Club gave Robert Bullard its top honor, the John Muir Award, and has established[24] an annual award to an individual or group that has done outstanding work in the area of environmental justice.

Linking Arms

As Afton showed us, communities of color acting jointly, even without significant wealth or political clout, have been far more effective than any number of individuals acting alone. And just as these first organizers did, don't forget to utilize preexisting groups if they are committed to environmental health or if you and your neighbors can inspire them to commit.

Here are steps to help you organize for the environmental health of your community:

Look Beyond Just Your Friends and Neighbors

Recruit your friends and neighbors, but also ask your church, sorority, and any other local groups to which you belong for their support.

As noted in Chapter 6, the EPA's "Guide for Community Partners," a blueprint for organizing, can be downloaded from https://nepis.epa.gov/Exe/ZyPDF.cgi?Dockey=P1004WLH.txt.

A website will help your group to share information and keep each other informed. It will also let others know what you are doing and about new issues as they arise. As a group, you will decide upon your goals for your community's health: Do you want a polluter's factory shut down? Do you want them to pay for cleanup? For testing of your children and homes? To establish a free clinic and medical bills resulting from the poisoning? To buy your homes sparing you the burden of trying to sell tainted property? Payment for damages? All these things have been achieved by communities just like yours. But you need a goal, a plan, and expert advice.

Don't be intimidated by the complexity of the work ahead of you because professionals with expertise in community organizing, law, and devising media campaigns are ready to help you. With such help, other groups have attained their goals, and you can too. Your church, temple, or other faith community can be a powerful ally to take on the dangers and inequities of environmental poisoning in your homes and neighborhoods, just as Afton's protest was strengthened by the Coley Springs Baptist Church and as the modern environmental racism movement was bolstered by the work of the United Church of Christ. Given the church's pivotal role as a seat of political and moral power behind every blow to violent racism, from abolition to desegregation to health disparities and environmental racism, the faith community has repeatedly shown itself to be an inspired agent of operation.

But it's not the only one.

Explore whatever organization you are active in to determine whether they are also potential partners for working to end the environmental poisoning of communities of color. Your sorority or fraternity, other social organization, your union, alumni group—whatever group in which you invest your effort, time, and heart—may be interested in pursuing health and social justice for marginalized communities: you'll never know until you ask.

The more you are able to tap into and mobilize existing organizations, the more efficient your efforts can become. Your group may find it helpful to read about other examples of environmental racism across the country, such as the ravaged communities described in this book. This may help you to understand the common actions and challenges as well as the strategies that polluters have embraced to evade detection, regulation, and justice.

It's important to learn that such communities are populated by middle-class as well as poor people of color, as in Anniston; that siting of such waste dumps is not always based on logical or scientific criteria as governments claim; that environmental waste and garbage services sometimes fail to provide minimal support as happened in Houston and Dunbar; that polluting industries or waste dumps were often imported without warning or acknowledgment of their hazards; that people are sometimes blamed by polluting industries for their deteriorating health as the Lead Institute blamed the "uneducable" parents of lead-poisoned black and Puerto Rican children, as municipal bureaucrats blamed parents of poisoned Baltimore children and later the water customers of Flint, Michigan, for poisonings that the city engineered.

Knowing this history can protect activists against succumbing to intimidation when experts intimate that poisoning woes are the fault of the victims or that the science dictating their actions is too complex for laypeople to understand.

Learn Where and How to Report Environmental Violations

Go to www.epa.gov/enforcement/report-environmental-violations. Complaints can be filed at www.epa.ie/enforcement/report/.

Hold Community Meetings

Post flyers throughout the neighborhood to invite everyone affected to the meetings and keep in touch by telephone/e-mail.

Write a Description of the Problem

You may have received health notices or read newspaper accounts of the poisoning that can help you with this. EJSCREEN, the environmental justice screening and mapping tool at https://www.epa.gov/ejscreen, can also be a very useful aid in characterizing and documenting the pollution problems in your community. It can also show you if neighboring communities are suffering from toxic exposures. If nearby communities have similar issues, try to join with them.

Align with Other Environmental Organizations

To find national environmental justice allies, use the map of Environmental Justice Health Alliance members who have freed themselves from environmental poisoning, often through coalitions of diverse organizations. The map is at http://ej4all.org/affiliates/. Join with organizations such as the Sierra Club and the Nature Conservancy, which work for stronger environmental and energy laws to reduce pollution and fossil-fuel dependence, to protect U.S. waterways and wildlife, and to preserve the wild. Their missions include keeping our air and water clean and replacing "dirty" fuel sources, ensuring a clean energy future. This makes their goals congruent with the struggles of those facing environmental racism.

Seek Legal Counsel

Before taking legal action, I strongly suggest that your group first contact a lawyer or Earth Justice, a legal agency that holds those who break our nation's laws accountable for their actions. Finding a legal ally or advocate is the best way to learn your legal rights and to decide which specific legal steps are most likely to bring you the results you want. If your group decides to file formal complaints or to sue, the paperwork involved can be complicated. Filing the wrong form and documents, completing them incorrectly, or filing them with the wrong agency can delay your relief. For free legal help in all this, as well as help in organizing around environmental dangers, consider consulting Earth Justice (https://earthjustice.org), which bills itself as "the nation's original and largest nonprofit environmental law organization."

This legal organization is currently pursuing approximately four hundred cases that focus on a variety of environmental concerns, including advising and offering free legal assistance to community groups who seek to end environmental assaults. (To obtain its detailed advice for community organizing, go to https://earthjustice.org/healthy-communities.) In order to end environmental injustice, you need to take the correct legal steps, warns Earth Justice, which offers free legal help around environmental concerns because, the group's website says, "The Earth needs a good lawyer." So does your community. You can reach Earth Justice at 800-584-6460 or via e-mail at info@earthjustice.org.

Make Your Concerns Known to Your Congressional Lawmakers

To find out who your senators and other elected representatives are, and how to contact them, go to https://earthjustice.org/action/faq#congress.

Take Advantage of Free Training

Look into the three-day Environmental Justice workshops conducted by the United Church of Christ. You can sign up at http://www.ucc.org/centers-for-environmental-justice. Representatives from your group should consider attending one; then they can take what they learn about developing community-based strategies and partnerships across ethnic and environmental interests back to your advocacy group, neighborhood, and city.

Apply for Grant Funding

Since 1994, the Environmental Justice Small Grants Program has awarded $24 million to more than 1,400 community-based organizations. For advice and/or funds to cover your group's expenses, contact the Office of Environmental Justice (OEJ) at https://www.epa.gov/environmentaljustice, especially "Environmental Justice Grants and Resources." Native American communities should contact the National Environmental Justice Advisory Council and Environmental Justice for Tribes and Indigenous Peoples at https://www.epa.gov/environmentaljustice/environmental justice-tribes-and-indigenous-peoples.

Devise a Media Strategy

After your organization decides upon its focus and goals, develop a media strategy to make more people aware of the problem. Recruiting an experienced media professional is crucial. Try these steps:

Visit the United Way (https://www.unitedway.org) for information on how to access volunteer media professionals.

Contact Encore Talent Works (https://toolkit.encore.org) for access to a media consultant.

Contact your local television news stations to explain your community's plight.

Write to your local print and online news outlet(s). For guidance on how to write an effective action letter, see https://earthjustice.org/action/faq#effective.

Challenges still confront those working to heal the poisoned world that sickens and confounds Americans, especially people of color. The concentration of toxic industrial wastes in ethnic neighborhoods remains—maintained by the polluters' callous denialism, clothed in the scientific lingo of "insufficient evidence," and joined by perennial "blame-the-victim" responses. This intellectual shrugging-off of mass poisoning is paralleled by other apostles of futility: namely, hereditarians who are wedded to the notion that innate, irredeemable racial deficiency, not the slow violence of widescale poisoning, impairs cognitive and behavioral functioning in America's ethnic enclaves.

But we know better. Problems stemming from brains injured by toxic exposures have solutions. From modest communities in Anniston, Dunbar, and Flint to mammoth ones like Baltimore, New York, and Washington, people of color and their white supporters understand that they must force government action to banish lead, mercury, industrial waste, and air pollution from cities, fence-line communities, Native American reservations, and "Cancer Alleys."

Many have traced the history of polluting U.S. corporations and shown they find it profitable to deny responsibility for death and disease, so we can't be surprised that industries still deem the costs of testing, cleanup, and abatement too great an expense to preserve the lives, health, and minds of millions.

Neither is the concentration of toxic killers and disablers among the racially marginalized shocking: it's happened too consistently to astonish us.

Moreover, the EPA has reached a nadir of malign neglect, signaled by Trump's appointment of Scott Pruitt, whose avowed aim was to dismantle the EPA. The rolling back of protections continued after his departure, hobbling efforts to provide potable water and curb emissions of toxics-spewing, mercury-producing coal-fired plants, while green-lighting questionable chemicals like chlorpyrifos. There's no denying that this diminution of our already feeble environmental-protection standards is discouraging.

So, what's the good news?

The odds seem stacked against those who militate for environmental justice, but this was precisely the case during the civil rights era, when activists persisted and won in the face of widespread murders and other unpunished violence, as well as in the face of hostile legal systems that protected segregation, lynching, unjust imprisonment, and marginalization. In rose-colored retrospect, it may appear as if the larger American public and news media championed civil rights workers, but many demonized them, calling for "law and order." Even the beloved Dr. Martin Luther King Jr. was, at the time of his death, widely derided as an immoral race-baiter and communist.

Yet he helped inspire the victorious civil rights warriors we now revere as heroes who saved the nation's justice system, and its soul. They would not stop until they won, and neither should we.

Today, activists of color and their white supporters in Flint, Anniston, Standing Rock, and smaller communities resist chemical oppression, just as the people of Afton refused to accept interstate transport of PCBs to poison their land and water.

Today, the activists do not battle corporate scientists and an indifferent government entirely on their own. The research and humanity of scientists like Herbert Needleman, Robert Bullard, Philip Landrigan, and Philippe Grandjean have armed them with a true picture of how industrial and environmental poisons damage minds as well as bodies. Lawyers like Suzanne Shapiro, Evan K. Thalenberg, and Thomas Yost have held corporations and institutions accountable for damage to the minds of lead-poisoned African American children.

And today a new generation of scientist-activists of color, such as Aaron Mair, the first African American director of the Sierra Club; Marsha Coleman-Adebayo, senior policy analyst for the EPA; and Mustafa Ali, cofounder of the EPA's environmental justice program, deploy their expertise and their passion for justice in this new civil rights movement that will win clean air, land, housing—and brains unimpaired by poisons.

Glossary

Abatement: Eliminating or dramatically reducing exposure to pollution.

Acute toxicity: Severe, relatively rapid biological harm or death caused after a single or brief toxic exposure.

Algae: Simple microscopic plants that grow in sunlit waters and provide food for fish and other marine animals.

Algal blooms: Sudden spurts of algal growth that can lower water quality or even poison it.

Asthma: A chronic, sometimes fatal disease of the lungs and airways that often has environmental triggers and that causes breathing difficulty.

Bisphenol A (BPA): An endocrine-disrupting chemical that either mimics or blocks hormones and disrupts the body's normal functioning.

Blood lead: The measurement of lead in one's blood, commonly measured with the metric **micrograms per deciliter, or μg/dL.** This is the number of micrograms of lead in a *tenth* of a liter of blood, where one microgram is roughly equivalent to 3.4 ounces. In other words, 1 μg/dL is equivalent to a grain of salt dissolved in 3.4 ounces of water.

Chlorination: The use of chlorine to disinfect drinking water, sewage, or wastes.

Chlorpyrifos: A chemical insecticide that prevents or destroys unwanted pests, such as cockroaches and mice.

Cognitive development: Brain growth from birth through adolescence that affects memory, speech and language, and problem-solving skills.

Cohort: A group of people in a research study.

Cumulative: Stores of some environmental hazards such as mercury and lead are additive or cumulative, accumulating in the bodies of food sources and people over years.

Emission: Pollutants released into the atmosphere from industrial facilities, residential chimneys, or from motor vehicle and aircraft exhaust.

Endocrine disruptors: Chemicals that interfere with the creation and deployment of natural hormones that govern intelligence and behavior, among other functions. Endocrine disruptors can hamper cognitive and brain development, causing learning disabilities, and any bodily system that is controlled by hormones can be harmed by these disruptors, which are found in many household and industrial products.

Environment: External substances and conditions that have effects on the life, development, and survival of an organism.

Environmental exposure: A person's contact with pollutants and the levels absorbed by their body.

Epigenetics: The study of environmental factors that alter gene expression.

Fecal coliform bacteria: Bacteria found in the intestinal tracts of mammals. Its presence in water or sludge is an indicator of pollution and possible contamination by pathogens.

GRAS: An FDA acronym for "Generally Recognized as Safe" that is applied to a food additive that has not undergone formal FDA review, but that, under FDA sections 201(s) and 409 of the Federal Food, Drug, and Cosmetic Act, has been deemed by qualified experts as "safe under the conditions of its intended use."[1]

Half-life: The time required for the amount of something to fall to half its initial value. For example, the half-life of DDT is about eight years, meaning that it takes about eight years for an animal to metabolize half the amount of DDT it absorbs.

Heavy metals: Metallic elements—for example, mercury and lead—that tend to accumulate in the food chain and damage living things.

IQ (intellectual quotient) tests: Standardized tests used to measure human intelligence. These tests are widely assumed to measure intellectual potential, but as Chapter 1 of this book notes, they do so in a limited manner that is fraught with bias, including racial bias. IQ tests do measure some skills and aspects of attainment and are correlated with measures of success such as income and longevity; again, less strongly in ethnic minority groups because of racial bias. The most common IQ tests include the Stanford-Binet Intelligence Scale, the Universal Nonverbal Intelligence Test, the Differential Ability Scales, the Peabody Individual Achievement Test, the Wechsler Individual Achievement Test, the Wechsler Adult Intelligence Scale, and the Woodcock-Johnson III Tests of Cognitive Disabilities.

Landfills: Disposal sites for waste. Some hold nontoxic material, but the most problematic landfills house wide varieties of hazardous wastes.

Micrograms per liter (μg/L): A unit of measurement for a concentration that expresses the mass of one material dissolved into the volume of another material. A microgram (μg) is one millionth of a gram, which is the mass of a single salt crystal. A liter is a metric unit of volume that is approximately equivalent to a quart. Also, 1 μg/L is equivalent to 1 part per billion (ppb). **Milligrams per deciliter (μg/dL)** is a related measurement that indicates the amount of a particular substance in a specific amount of blood.

Naphthalene: A class of hydrocarbon compounds used as pesticide. Found in mothballs and air pollution.

Organophosphates: Phosphorus-containing pesticides used to kill insects.

Parts per billion (ppb); parts per million (ppm): Metric units used to express exceedingly small amounts of potent contaminants in air, water, soil, human tissue, food, and/or other products.

A simple illustration of these quantities:

Unit	1 ppm	1 ppb
Weight	1 ounce in 31 tons	1 pinch of salt in 10 tons
Area	1 square foot in 23 acres	1 square foot in 36 square miles
Concentration	———	1 µg/L equals 1 salt crystal in 3.4 ounces (see **"Micrograms per liter (µg/L)"** above)

Pathogen: Any microorganism—for example, an alga, bacterium, virus, or fungus—that causes disease in humans.

Pesticides: Substances used to repel or destroy any pest. Pesticides can accumulate in the food chain and contaminate the environment.

Phenols: Organic compounds that are by-products of petroleum refining, tanning, and resin manufacturing. Low concentrations create foul tastes and odors in water, but higher concentrations can kill aquatic and human life.

Phosphates: Certain chemical phosphorus compounds.

Phthalates: These chemicals, which are used to soften many plastics in household, kitchen, and cosmetic products, can seep out and disrupt the body's hormone regulation by affecting the endocrine system.[2]

Poison: Any substance that can cause illness or death when administered or absorbed.

Polybrominated diphenyl ethers (PBDEs): Widely used as flame retardants and in building materials, they have been shown to reduce fertility in humans at levels found in households.

Polycyclic aromatic hydrocarbons (PAH): Tiny cancer-causing particles of pollution that result from the burning of fuel.

Pyrethroids: Most commercial household insecticides now contain these chemicals, which are also used for insulating purposes and in gas pipeline systems as lubricants.

Racial groups: The nomenclature of U.S. racial groups is imperfect and in some cases hotly contested. Because race is a social construct that maps poorly onto genetics, my goal is a social one as well: I intend to call people what they wish to be called. After conferring with experts who are members of various racial communities, I have chosen what appears to be the most accurate, least offensive terminology. *African American* refers to any U.S. resident who identifies as a person of African descent. I have used *black* when it is the terminology of the research study I am discussing; some such studies may include people other than Americans. *Native American* broadly refers to the indigenous peoples of the United States and its territories, although where possible I have used specific kinship designations. My data sources rarely do this, however, so *Native American* is most frequently used. *Latino* and *Hispanic* are often treated as synonymous, although, as I mention in the Introduction, the former refers to land of origin and the latter to language. I have eschewed the neologism *LatinX* because there is not yet a consensus regarding its use among the communities it describes. As I explain in the Introduction, Hispanic and Latino people can be of any race.

Asians are broadly designated in much of the research I consulted, so this book tends to use the term "Asian" rather than using specific nationalities or kinship groups. "White" is used to refer to people who do not identify with another ethnic group, such as Hispanic, Asian, or African American. In the Introduction, I discuss how the medical literature deploys these racial categories in an illogical, inconsistent, and inaccurate manner.

Red tide: A proliferation of a marine plankton that is toxic and often fatal to fish and sometimes, as Chapter 5 notes, to the humans who eat them as well. A tide can be called red, green, or brown, depending on the color of the plankton.

Risk assessment: A qualitative and quantitative evaluation performed in an effort to define the risk posed to human health or the environment by the use and presence or potential presence of specific pollutants.

Sludge: The semisolid residue left after air or water treatment. It can be hazardous waste.

Social stressors: Environmental conditions such as poverty, racial bias, high rates of disease, overcrowding, high rates of violent crime, unemployment, and substandard housing conditions, all of which affect most people of color in environmentally blighted communities.

Teratogen: A drug, agent, or other influence that causes physical defects in a developing embryo.

Toxic: Another term for *poisonous.*

Toxicant: A type of poison made by humans. This is in contrast to a toxin, which is a protein-based poison generated naturally by a plant or an animal.

Toxin: A poisonous substance produced within living cells or organisms. Thus, most industrial and man-made chemicals can be called *toxic* but cannot be described as *toxins.*

Trichloroethylene (TCE): Toxic liquid used as a solvent, a metal-degreasing agent, and in other industrial applications, notably dry cleaning.

Vector: An organism, often an insect or rodent, that carries disease.

Volatile Organic Compounds (VOCs): Common pungent organic solvents that are frequently toxic and easily vaporize at room temperature and under normal pressure. VOCs include gasoline, kerosene, dry-cleaning products, aerosol spray, can propellants, and many other familiar products.

List of Known Chemical Brain Drainers

On the website for his book *Only One Chance: How Environmental Pollution Impairs Brain Development—and How to Protect the Brains of the Next Generation,* Philippe Grandjean, head of the Environmental Medicine Research Unit at the University of Southern Denmark and adjunct professor of environmental health at the Harvard School of Public Health, provides the following list of more than two hundred industrial chemicals that are toxic to the brain. These often cause neurological symptoms. The list is not meant to be exhaustive. Updates, along with Grandjean's discussions of new developments, can be found at http://braindrain.dk/known-chemical-brain-drainers. All chemicals marked with an asterisk are high-production-volume substances in both the United States and Europe.

Metals and Inorganic Compounds

Aluminum
Arsenic and arsenic compounds
Azides
Barium compounds
Bismuth compounds
Carbon monoxide*
Cyanides
Decaborane
Diborane
Ethylmercury
Fluorides
Hydrogen phosphide
Hydrogen sulfide*
Lead and lead compounds*
Lithium compounds
Manganese and manganese compounds*

Mercury and mercuric compounds
Methylmercury
Nickel carbonyl
Pentaborane
Phosphine
Phosphorus*
Selenium compounds
Tellurium compounds
Thallium compounds
Tin compounds*

Pesticides

Acetamiprid
Aldicarb (Temik)
Aldrin
Amitraz
Avermectin
Bensulide
Bromophos (Brofene)
Carbaryl (Sevin)*
Carbofuran (Furadan)*
Carbophenothion (Trithion)
α-Chloralose
Chlordane
Chlorfenvinphos
Chlormephos
Chlorothion
Chlorpyrifos (Dursban, Lorsban)*
Coumaphos
Cyhalothrin (Karate)
Cyolane (Phospholan)
Cypermethrin
Deltamethrin (Decamethrin)
Demeton-S-methyl
Dialifor
Diazinon*
1,2-Dibromo-3-chloropropane (DBCP)
Dichlofenthion
Dichlorodiphenyltrichloroethane (DDT)*
2,4-Dichlorophenoxyacetic acid (2,4-D)*
1,3-Dichloropropene*
Dichlorvos (DDVP, Vapona)
Dieldrin
Dimefox
Dimethoate*
4,6-Dinitro-o-cresol
Dinoseb*
Dioxathion
Disulfoton
Edifenphos
Emamectin
Endosulfan (Thiodan)*
Endothion
Endrin
Ethiofencarb (Croneton)
Ethion
Ethoprop

O-Ethyl O-(4-nitrophenyl) phenylphosphonothioate (EPN)
Fenitrothion
Fensulfothion
Fenthion*
Fenvalerate
Fipronil (Termidor)
Fonofos
Formothion
Glyphosate
Heptachlor
Heptenophos
Hexachlorobenzene*
Hexaconazole
Imidacloprid
Isobenzan
Isolan
Isoxathion
Kepone (Chlordecone)
Leptophos
Lindane (γ-Hexachloro cyclohexane)*
Merphos*
Metaldehyde*
Methamidophos*
Methidathion (Suprathion)
Methomyl
Methyl bromide*
Methyl parathion (Parathion-methyl)*
Mevinphos
Mexacarbate (Zectran)
Mipafox
Mirex
Monocrotophos
Naled
Nicotine
Oxydemeton-methyl
Parathion*
Pentachlorophenol
Phorate
Phosphamidon (Dimecron)
Propaphos
Propoxur (Baygon)*
Pyriminil (Pyrinuron, Vacor)
Sarin
Schradan
Soman
Sulprofos
Systox (Demeton)
Tebupirimfos
Tefluthrin
Terbufos
Tetramethylenedisulfotetramine (Tetramine)
Thiram*
Toxaphene
Trichlorfon
Trichloronate
2,4,5-Trichlorophenoxyacetic acid (2,4,5-T)

Organic Solvents

Acetone*
Benzene*
Benzyl alcohol*
1-Bromopropane*
Carbon disulfide*
Chloroform*
Cyclohexane*
Cyclohexanol*
Cyclohexanone*
1,2-Dibromoethane*
Dichloroacetic acid
Dichloromethane*
Diethylene glycol*
N,N-Dimethylformamide*
Ethanol (alcohol)*
Ethyl acetate*
Ethyl chloride
Ethylene glycol*
Ethylene glycol ethyl ether (Ethoxyethanol)*
Ethylene glycol methyl ether (Methoxyethanol or Methyl cellosolve)*
n-Hexane*
Isophorone*
Isopropyl alcohol*
Methanol (Methyl alcohol)*
Methylcyclopentane
Methyl ethyl ketone*
Methyl isobutyl ketone*
Methyl-n-butyl ketone (2-Hexanone)
2-Methylpropanenitrile*
Nitrobenzene*
2-Nitropropane*
1-Pentanol*
Pyridine*
Styrene*
1,1,2,2-Tetrachloroethane*
Tetrachloroethylene (Perchloroethylene)*
Toluene*
1,1,1-Trichloroethane (Methylchloroform)*
Trichloroethylene*
Xylenes*

Other Organic Substances

Acetone cyanohydrin (2-Hydroxy-2-methylpropanenitrile)*
Acrylamide (2-Propenamide)*
Acrylonitrile*
Allyl chloride (1-Chloro-2-propene)*
Aniline*

1,4-Benzenediamine (4-Aminoaniline)*
1,2-Benzenedicarbonitrile (1,2-Dicyanobenzene)*
Benzonitrile*
1,3-Butadiene*
Butylated triphenyl phosphate*
Caprolactam (Azepan-2-one)*
Chloroprene*
Cumene*
Cyclonite (RDX)*
Diethylene glycol diacrylate*
3-(Dimethylamino)-propanenitrile*
Dimethylhydrazine*
Dimethyl sulfate*
Di-N-butyl phthalate*
Dinitrobenzene*
Dinitrotoluene*
Ethylbis(2-chloroethyl) amine
Ethylene
Ethylene oxide*
Fluoroacetamide
Fluoroacetate*
Hexachlorophene
Hydrazine*
Hydroquinone*
Methyl chloride (Chloromethane)*
Methyl formate*
Methyl iodide
Methyl methacrylate*
4-Nitroaniline*
Phenol*
Phenylhydrazine
Polybrominated biphenyls (PBBs)
Polybrominated diphenyl ethers (PBDEs)
Polychlorinated biphenyls (PCBs)
1,2-Propylene oxide*
2,3,7,8-Tetrachloro dibenzo-p-dioxin (TCDD)
Tributyl phosphate*
Trimethyl phosphate
Tri-o-cresylphosphate
Triphenyl phosphate*
Tris(2-chloroethyl)amine (Trichlormethine)
Vinyl chloride (Chloroethene)*

Acknowledgments

My focus on the mental and intellectual devastation wrought by the exposure of marginalized racial groups to toxic environmental contaminants is novel, but my reliance on the work of many productive scientists is not.

I remain especially grateful to those who, from this work's inception, were generous with their expertise, resources, and valuable time. These include Robert Bullard, founder of the EPA Office of Environmental Equity and "grandfather" of the environmental justice movement; University of New Mexico evolutionary biology professor Randy Thornhill; Carl Bell, director of the Institute for Juvenile Research at the University of Illinois at Chicago; and Gerald Markowitz, coauthor with David Rosen of the incisive *Lead Wars: The Politics of Science and the Fate of America's Children.*

Decades ago, working at the Lifeline/Finger Lakes Regional Poison Control Center under the supervision of its founder, Ruth A. Lawrence—a neonatologist, a clinical toxicologist, and author of *Breastfeeding: A Guide for the Medical Profession*—made me aware of the terrible versatility of toxic exposures and their effects on the mind as well as the body, and also of the special vulnerabilities of children, infants, and the unborn. I addressed these in my 2015 book, *Infectious Madness,* and in "The Well Curve," an article published the same year in *The American Scholar.* Professor Rosen's class Poisoned World, at Columbia University, inspired me to think ever more broadly about the sea of toxicity in which too many American communities are steeped.

Revelatory publications and conversations with Christopher Eppig, director of programing for the Chicago Council on Science and Technology, were invaluable, as was the work of Dorothy Roberts, University of Pennsylvania George A. Weiss University professor of law and sociology and author of *Fatal Invention: How Science, Politics, and Big Business Re-Create Race in the Twenty-First Century*. University of Alberta professor Kim TallBear, author of *Native American DNA: Tribal Belonging and the False Promise of Genetic Science* also helped inform my discussions of race, intelligence, and IQ. So did my exchanges with Girma Berhanu, professor of education at the University of Gothenburg, Sweden, and with Brink Lindsey, author of *Human Capitalism: How Economic Growth Has Made Us Smarter—and More Unequal.*

Developmental toxicologist Philip Landrigan, Bruce Lanphear, Robert Yolken, and Peter Hotez generously answered my many questions about heavy metals, pathogens, and other environmental assaults. Philippe Grandjean, head of the Environmental Medicine Research Unit at the University of Southern Denmark and adjunct professor of environmental health at Harvard's T. H. Chan School of Public Health, was not only generous with his time, but also allowed me to reproduce his published data, including his list of chemical "brain drainers."

The dynamic duo of Betsy and Robin Lindsey—the features editors for the History News Network—offered early and welcome support. So did Nathan Lents, John Jay College professor of biology, and University of Washington School of Dentistry professor Donald Chi.

This book could not have been written without the dogged courage of those who resisted the poisoning of their children and communities in the face of sickness, premature death, "blame the victim" demonization, and even incarceration. Thank you, David Baker, Shirley Carter, Tara Adams, Maritza Lopez, and all those who have braved indifference from the very medical and governmental systems that should protect them.

Thank you, too, to the legal champions who have secured and still seek justice for them. Among these are Baltimore lawyers Evan K. Thalenberg, Saul Kerpelman, Suzanne Shapiro, Thomas F. Yost, and Nicholas Szokoly. As a consultant for them, I have witnessed firsthand the skill and dedication with which they procure that justice.

I'm deeply indebted to Arlene Shaner, historical collections librarian at the New York Academy of Medicine, for finding reproducible images of century-old illustrations after I had all but given up hope, and for going the extra mile by scanning high-resolution versions, on deadline. Tricia Gesner of the Associated Press also helped me obtain elusive images.

I am grateful as always for the warmth and keen judgment of my wonderful agent, Lisa Bankoff, and I am thrilled by the opportunity to thank my editor, Tracy Behar, vice president, publisher, and editor-in-chief of Little, Brown Spark for her insight and organizational genius; and Ian Straus, for his close attention and support.

I also thank the rest of Little, Brown team, including Lena Little, Ben Allen, and Peggy Freudenthal. And I'm grateful to Chris Nolan for the kindly laser of his legal review of the manuscript and the generosity with which he dispensed advice.

I thank Alondra Nelson, social science chair at the Institute for Advanced Study; Hilda Hutcherson, professor and associate dean of Columbia University Medical Center; the *New York Times*'s Sheri L. Fink; and Charles Ezra Ferrell, visionary vice president of Detroit's Charles H. Wright Museum of African American History—all of whom gave me early and constant support. So did Joshua Prager, Randi Hutter Epstein, and my fellow scribes of the Invisible Institute, founded by Annie Murphy Paul and Alissa Quart, who have proved to be trusted advisers. These pillars of support include Kaja Perina, Susan Cain, Abby Ellin, Tom Zoellner, Wendy Paris, Christine Kenneally, Catherine Orenstein, Katherine Stewart, Elizabeth DeVita-Raeburn, Paul

Raeburn, Maia Szalavitz, Stacy Sullivan, Gretchen Rubin, Judith Matloff, Lauren Sandler, Ada Calhoun, Gary Bass, Bob Sullivan, and Ron Lieber

Last but definitely not least, I am blessed with Donna and Tom Harman for their brilliance and kindness, and with Kate, Eric, and Theresa, my sisters and brother. And I miss Pete every day.

Ron DeBose is always with me.

Notes

Introduction: IQ Matters

1. William Shockley, “Models, Mathematics, and the Moral Obligation to Diagnose the Origin of Negro IQ Deficits,” *Review of Educational Research* 41, no. 4 (1971).
2. R. E. Nisbett, *Intelligence and How to Get It: Why Schools and Cultures Count* (New York: Norton, 2009), 93.
3. James McCune Smith, M.D., “The Memorial of 1844 to the U.S. Senate” (1844), in Herbert M. Morais, *The History of the Afro-American in Medicine,* rev ed. (Cornwell Heights, PA: The Publishers Agency, 1978), 212–13; *New York Tribune,* May 5, 1844, and *The Liberator,* May 31, 1844; James McCune Smith, “Facts Concerning Free Negroes,” in *A Documentary History of the Negro People in the United States,* ed. Herbert Aptheker (New York: Citadel, 1844); W. Montague Cobb, “The Negro as a Biological Element in the American Population,” *Journal of Negro Education* 8, no. 3 (1939): 336–48.
4. Stephen Jay Gould, *The Mismeasure of Man* (New York: Norton, 1981); Robert V. Guthrie, *Even the Rat Was White: A Historical View of Psychology* (1976; repr., Needham, MA: Allyn & Bacon, 1998).
5. James Feyrer, Dimitra Politi, and David N. Weil, “The Cognitive Effects of Micronutrient Deficiency: Evidence from Salt Iodization in the United States,” National Bureau of Economic Research Working Paper, July 18, 2013.
6. E. F. Eldridge, “The Iodine Content of Michigan Water Supplies,” *The American Journal of Public Health* 14, no. 9 (September 1924): 750–54.
7. Quoted in the Salt Institute, “Iodized Salt,” July 13, 2013, http://www.saltinstitute.org/2013/07/13/iodized-salt/.
8. Philippe Grandjean, *Only One Chance,* 1st ed. (New York: Oxford University Press, 2013); see also “A Ted Wells tribute to Johnnie Cochran with a ‘30 Frames’ edge,” https://www.youtube.com/watch?v=UyUpye-ViiE.
9. Tom Neltner, “Federal Court of Appeals Gives EPA One Year to Update Lead-Based Paint Standards,” Environmental Defense Fund, February 3, 2017.
10. Ibid.

11. Vann R. Newkirk II, "Fighting Environmental Racism in North Carolina," *The New Yorker,* January 16, 2016, https://www.newyorker.com/news/news-desk/fighting-environmental-racism-in-north-carolina.
12. Philippe Grandjean, "Known Chemical Brain Drainers," http://braindrain.dk/known-chemical-brain-drainers/.
13. Robert Bullard, telephone interview with the author, October 27, 2016.
14. Simon Mahan and Kimberly Warner, "Hidden Costs: Reduced IQ from Chlor-Alkali Plant Mercury Emissions Harm the Economy," *Oceana,* May 2009, https://usa.oceana.org/sites/default/files/reports/Hidden_Costs.pdf.
15. Andrew Curry, "Why Living in a Poor Neighborhood Can Change Your Biology," *Nautilus,* June 14, 2018, https://getpocket.com/explore/item/why-living-in-a-poor-neighborhood-can-change-your-biology.
16. The Associated Press, "Trump EPA Orders Rollback of Obama Mercury Regulations," reprinted *New York Times,* December 28, 2018, https://www.nytimes.com/aponline/2018/12/28/us/politics/ap-us-clean-air-mercury.html.
17. Lynn Peeples, "'Little Things Matter' Exposes Big Threat to Children's Brains," *Huffington Post,* November 20, 2014, updated December 6, 2017; https://www.huffingtonpost.com/2014/11/20/toxins-children-brain-little-things-matter_n_6189726.html.
18. Author's telephone interview with Brink Lindsey, June 15, 2016.
19. "Cognitive ability was more than five times more powerful than emotional intelligence. The average employee with high cognitive ability generated annual revenue of over $195,000, compared with $159,000 for those with moderate cognitive ability and $109,000 for those with low cognitive ability. Emotional intelligence added nothing after measuring cognitive ability." Adam Grant, "Emotional Intelligence Is Overrated," LinkedIn, September 30, 2014, https://www.linkedin.com/pulse/20140930125543-69244073-emotional-intelligence-is-overrated; "Adam M. Grant Rocking the Boat but Keeping It Steady: The Role of Emotion Regulation in Employee Voice," *Academy of Management Journal* 56, no. 6 (November 30, 2017): 1,703–23, https://journals.aom.org/doi/abs/10.5465/amj.2011.0035. See also, Tomas Chamorro-Premuzic, Seymour Adler, and Robert B. Kaiser, "What Science Says about Identifying High-Potential Employees," *Harvard Business Review,* October 3, 2017, https://hbr.org/2017/10/what-science-says-about-identifying-high-potential-employees.
20. Kimberly Heffling, "US Adults Score Below Average on Worldwide Competency Test," Associated Press, October 8, 2013, https://www.businessinsider.com/us-adults-score-below-average-on-worldwide-competency-test-2013-10.
21. Cited in ibid.
22. Nik Ahmad Sufian Burhan, Mohd Rosli Mohamad, Yohan Kurniawan, and Abdul Halim Sidek, "The Impact of Low, Average, and High IQ on Economic Growth and Technological Progress: Do All Individuals Con-

tribute Equally?," April 15, 2014, *Munich Personal RePEc Archive,* MPRA Paper no. 77321, https://mpra.ub.uni-muenchen.de/77321/.

Chapter 1: The Prism of Race: How Politics Shroud the Truth about Our Nation's IQ

1. Stephen Jay Gould, *The Panda's Thumb: More Reflections in Natural History* (New York: Norton, 2010).
2. Susan Dynarski, "Why Talented Black and Hispanic Students Can Go Undiscovered," *New York Times,* April 8, 2016, https://www.nytimes.com/2016/04/10/upshot/why-talented-black-and-hispanic-students-can-go-undiscovered.html.
3. Author's telephone interview with Christopher Eppig, June 15, 2016.
4. Brink Lindsey, *Human Capitalism: How Economic Growth Has Made Us Smarter–and More Unequal,* rev. ed. (Princeton, NJ: Princeton University Press, 2013); author's interview with Brink Lindsey, June 15, 2016.
5. Arthur R. Jensen, "How Much Can We Boost IQ and Scholastic Achievement?," *Harvard Educational Review,* 39 (1969): 1–123.
6. Ibid., 5.
7. Thomas Volken, "The Impact of National IQ on Income and Growth: A Critique of Richard Lynn and Tatu Vanhanen's Recent Book," January 2005, https://web.archive.org/web/20120229125855/http://www.history.ox.ac.uk/hsmt/courses_reading/undergraduate/authority_of_nature/week_8/volken.pdf.
8. Girma Berhanu, "Black Intellectual Genocide: An Essay Review of *IQ and the Wealth of Nations,*" *Education Review* 10(6) (2007): 1–28.
9. Brink Lindsey, "Why People Keep Misunderstanding the 'Connection' between Race and IQ," *The Atlantic,* May 15, 2013, https://www.cato.org/publications/commentary/why-people-keep-misunderstanding-connection-between-race-iq.
10. Girma Berhanu, "Academic Racism: Lynn's and Kanazawa's Ill-Considered Theory of Racial Differences in Intelligence," *Education Review* 14, no. 12 (2011): 3–4; also, e-mail exchange between Girma Berhanu and the author, June 19, 2015.
11. Berhanu, "Academic Racism," 3–4; Girma Berhanu, "Black Intellectual Genocide"; also, e-mail exchange between Girma Berhanu and the author, June 19, 2015.
12. L. Eyferth, "Performance of Different Groups of Children of Occupation Forces on the Hamburg-Wechsler Intelligence Test for Children," *Archiv für die gesamte Psychologie* 113 (1961): 222–41.
13. Eric Turkheimer, Kathryn Paige Harden, and Richard E. Nisbett, "There's Still No Good Reason to Believe Black-White IQ Differences Are Due to Genes," *Vox,* June 17, 2017, https://www.vox.com/the-big-idea/2017/6/15/15797120/race-black-white-iq-response-critics?fbclid=IwAR3oTAbQVKSEd6t_1AQyLc2cMZAacRBPwNlW8LtY1P1_3aBG96yoFGDtT2E.

14. L. Eyferth, "Performance of Different Groups of Children of Occupation Forces on the Hamburg-Wechsler Intelligence Test for Children."
15. "Flynn's study revealed a 13.8-point increase in IQ scores between 1932 and 1978, amounting to a 0.3-point increase per year, or approximately 3 points per decade. More recently, the Flynn effect was supported by calculations of IQ score gains between 1972 and 2006 for different normative versions of the Stanford-Binet (SB), Wechsler Adult Intelligence Scale (WAIS), and Wechsler Intelligence Scale for Children (WISC) (Flynn, 2009a). The average increase in IQ scores per year was 0.31, which was consistent with Flynn's (1984a) earlier findings." Lisa Trahan, "The Flynn Effect: A Meta-analysis," *Psychological Bulletin* 140, no. 5 (September 2014): 1,332–60, doi: 10.1037/a0037173.
16. Harriet A. Washington, *Medical Apartheid: The Dark History of Medical Experimentation on Black Americans from Colonial Times to the Present* (New York: Random House, 2007).
17. Ibid., especially chapter 2; Josiah Nott, "The Mulatto a Hybrid—Probable Extermination of the Two Races If the Whites and Blacks Are Allowed to Intermarry," *American Journal of the Medical Sciences* 6 (1843); Lathan A. Windley, "Runaway Slave Advertisements of George Washington and Thomas Jefferson," *Journal of Negro History* 63, no. 4 (1978): 373–74; "Eighteenth Century Slaves as Advertised by Their Masters," *Journal of Negro History* 1, no. 2 (1916): 163–216; Eric L. McKitrick, *Slavery Defended: The Views of the Old South* (Englewood Cliffs, NJ: Prentice-Hall, 1963); James McCune Smith, "Facts Concerning Free Negroes," in *A Documentary History of the Negro People in the United States,* ed. Herbert Aptheker (New York: Citadel, 1844); W. Montague Cobb, "The Negro as a Biological Element in the American Population," *Journal of Negro Education* 8, no. 3 (1939): 336–48.
18. Samuel George Morton, *Crania Americana; Or a Comparative View of the Skulls of Various Aboriginal Nations of North and South America: To Which Is Prefixed an Essay on the... Human Species at Species (Classic Reprint)* (London: Forgotten Books, 2012); see also Vassar College Archaeology blog, "Phrenology and 'Scientific Racism' in the 19th Century," March 5, 2017, https://pages.vassar.edu/realarchaeology/2017/03/05/phrenology-and-scientific-racism-in-the-19th-century/; also see Stephen Jay Gould, *The Mismeasure of Man* (New York: Norton, 1981).
19. Quoted in Harriet A. Washington, *Medical Apartheid: The Dark History of Medical Experimentation on Black Americans from Colonial Times to the Present* (New York: Random House, 2007), 36.
20. Gould, *The Mismeasure of Man;* Robert V. Guthrie, *Even the Rat Was White: A Historical View of Psychology* (1976; repr. Needham, MA: Allyn & Bacon, 1998).
21. James McCune Smith, M.D., "The Memorial of 1844 to the U.S. Senate" (1844), in Herbert M. Morais, *The History of the Afro-American in Medicine,* rev ed. (Cornwell Heights, PA: The Publishers Agency, 1978), 212–13.

22. M. R. Trabue, "The Intelligence of Negro Recruits," *Natural History* 19, no. 6 (December 1919): 685.
23. "The Development of the [Army] beta test and of the performance test for the examination of the foreign speaking and illiterate presented special problems. The use of demonstration charts and mime to convey the instructions to the persons being examined proved successful. The new type of the test in the beta, using geometrical designs, mutilated pictures, etc., required different principles in its construction. The individual performance tests also involved additional and peculiar standards of construction and evaluation." Clarence Stone Yoakum and Robert Mearns Yerkes, *Army Mental Tests* (New York: Henry Holt and Company, 1920).
24. W. E. B. Du Bois, "Race Intelligence," *The Crisis* 20, no. 3 (July 1920): 326; http://www.webdubois.org/dbRaceIntell.html; also see Trabue, "The Intelligence of Negro Recruits," 685.
25. Cited in David Weigel, "The IQ Test," *Slate*, May 10, 2013, http://www.slate.com/articles/news_and_politics/politics/2013/05/jason_richwine_hispanics_and_iqs_the_heritage_foundation_scholar_began_researching.html.
26. Pilar Ossorio, "Race, Genes and Intelligence," *Genewatch* 22, nos. 3–4 (2009): n.p., http://www.councilforresponsiblegenetics.org/Projects/CurrentProject.aspx?projectId=8.
27. Ibid.
28. Robert Anemone, "The Hereditarian Theory of IQ: An American Invention," Seminar in Biological Anthropology blog, October 21, 2009, http://anthropology6030.blogspot.com/2009/10/hereditarian-theory-of-iq-american.html.
29. M. J. Kane, D. Z. Hambrick, and A. R. A. Conway, "Working memory capacity and fluid intelligence are strongly related constructs," *Psychological Bulletin* 131 (2005): 66–71.
30. Author's interview with Brink Lindsey, June 15, 2016.
31. J. L. Kincheloe, S. R. Steinberg, and A. D. Gresson, eds., *Measured Lies* (New York: St. Martin's Press, 1996); C. S. Fisher et al., eds., *Inequality by Design: Cracking the Bell Curve Myth* (Princeton, NJ: Princeton University Press, 1996); M. K. Brown et al., *Whitewashing Race: The Myth of a Colorblind Society* (Berkeley: University of California Press, 2003); R. E. Nisbett, *Intelligence and How to Get It: Why Schools and Cultures Count* (New York: Norton, 2009).
32. R. Lynn and T. Vanhanen, *IQ and the Wealth of Nations* (Westport, CT: Praeger, 2002); Berhanu, "Academic Racism," 3–4.
33. Berhanu, "Academic Racism," 3–4.
34. S. Kanazawa, "Temperature and Evolutionary Novelty as Forces Behind the Evolution of General Intelligence," *Intelligence* 36 (2008): 107.
35. S. Kanazawa, "Mind the Gap...in Intelligence: Re-examining the Relationship between Inequality and Health," *British Journal of Health Psychology* 11

(2006): 623–42; also see S. Kanazawa, "Temperature and Evolutionary Novelty as Forces Behind the Evolution of General Intelligence," 99–108.

36. Mikhail Lyubansky, "Kanazawa Apologizes for 'Black Unattractiveness' Article, Apparently Gets to Keep Job," *Psychology Today,* September 11, 2011, https://www.psychologytoday.com/blog/between-the-lines/201109/kanazawa-apologizes-black-unattractiveness-article-apparently-gets; see also Satoshi Kanazawa, "Why Are Black Women Less Physically Attractive Than Other Women?," http://tishushu.tumblr.com/post/5548905092/here-is-the-psychology-today-article-by-Satoshi-Kanazawa.
37. Wolfgang Saxon, "William B. Shockley, 79, Creator of Transistor and Theory on Race," *New York Times,* August 14, 1989; "Shockley's Thesis," *Firing Line with William F. Buckley Jr.,* June 10, 1974, https://www.youtube.com/watch?v=7JOIqkh2ms8; Scott Rosenberg, "Silicon Valley's First Founder Was Its Worst," *Wired,* July 19, 2017, https://www.wired.com/story/silicon-valleys-first-founder-was-its-worst/.
38. "Shockley's Thesis."
39. Cahal Milmo, "Fury at DNA Pioneer's Theory: Africans Are Less Intelligent Than Westerners," *The Independent,* October 16, 2007, http://www.independent.co.uk/news/science/fury-at-dna-pioneers-theory-africans-are-less-intelligent-than-westerners-394898.html.
40. Roxanne Khamsi, "James Watson Retires Amidst Race Controversy," *Daily News,* October 25, 2007.
41. Charles Lane, "Daniel R. Vining Jr., reply by Charles Lane," *New York Review of Books* 42, no. 5 (March 23, 1995): https://www.nybooks.com/articles/1995/03/23/pioneer/.
42. Andrew Duffy, "Rushton Revisited," *The Ottawa Citizen,* October 1, 2005, A1, http://lists.extropy.org/pipermail/paleopsych/2005-October/004309.html.
43. Zack Cernovsky, Ph.D., discussion with the author, New Orleans, May 24, 2010.
44. Graham Coop et al., "Letters: 'A Troublesome Inheritance,'" *New York Times,* August 8, 2014, https://www.nytimes.com/2014/08/10/books/review/letters-a-troublesome-inheritance.html.
45. Sharon Begley, "Three Is Not Enough," *Newsweek,* February 12, 1995, http://www.newsweek.com/three-not-enough-184974.
46. Barbara Chase-Riboud, *The President's Daughter: A Novel* (Rediscovered Classics) (Chicago: Chicago Review Press, 2009).
47. Catherine Kerrison, *Jefferson's Daughters: Three Sisters, White and Black, in a Young America* (New York: Ballantine, 2018); Catherine Kerrison, "Commentary: How Did We Lose a President's Daughter?," *Washington Post,* January 27, 2018, http://www.bendbulletin.com/opinion/5950137-151/commentary-how-did-we-lose-a-presidents-daughter; also see Frank W. Sweet's *A Legal History of the Color Line: The Rise and Triumph of the One-Drop Rule* (Palm Coast, FL: Backintyme Publishers, 2013).
48. Kerrison, "Commentary: How Did We Lose a President's Daughter?"
49. Sweet, *A Legal History of the Color Line.*

50. Begley, "Three Is Not Enough."
51. Ossorio, "Race, Genes and Intelligence."
52. Cited in J. Patrick Coolican, "'Myth of Model Minority' Targeted," *Seattle Times,* April 24, 2003.
53. Robert Rosenthal and Lenore Jacobson, *Pygmalion in the Classroom: Teacher Expectation and Pupils' Intellectual Development,* rev. ed. (Carmarthen, Wales: Crown House Publishing, 1992), *passim.*
54. Scott Jaschik, "Anti-Asian Bias Claim Rejected," *Inside Higher Ed,* September 24, 2015.
55. Anemona Hartocollis, "Does Harvard Admissions Discriminate? The Lawsuit on Affirmative Action, Explained," *New York Times,* October 15, 2018, https://www.nytimes.com/2018/10/15/us/harvard-affirmative-action-asian-americans.html.
56. Michael Li, "I Support Affirmative Action. But Harvard Really Is Hurting Asian Americans," *Vox,* October 18, 2018, https://www.vox.com/first-person/2018/10/18/17995270/asian-americans-affirmative-action-harvard-admissions-lawsuit.
57. Adam Liptak, "Supreme Court Upholds Affirmative Action Program at University of Texas," *New York Times,* June 23, 2016, https://www.nytimes.com/2016/06/24/us/politics/supreme-court-affirmative-action-university-of-texas.html?module=inline.
58. Emil Guillermo, "The Biggest Diversity Case in 2018 Could Be the Biggest of 2019," *Diverse Issues in Higher Education,* December 28, 2018, https://diverseeducation.com/article/134999/.
59. Frank Wu, *Yellow: Race in America Beyond Black and White* (New York: Basic Books, 2003), 67.
60. Judkin Browning, *Shifting Loyalties: The Union Occupation of Eastern North Carolina* (Raleigh, NC: The University of North Carolina Press, 2014), 129.
61. Alec MacGillis, "The Original Underclass," *The Atlantic,* September 2016.

Chapter 2: The Lead Age: Heavy Metals, Low IQs

1. Abby Goodnough, "Flint Weighs Scope of Harm to Children Caused by Lead in Water," *New York Times,* January 29, 2016.
2. David Rosner and Gerald Markowitz, *Lead Wars: The Politics of Science and the Fate of America's Children* (Berkeley: University of California Press, 2014), 132.
3. Harriet A. Washington, *Medical Apartheid: The Dark History of Medical Experimentation on Black Americans from Colonial Times to the Present* (New York: Doubleday, 2007), 291.
4. Ibid., 292.
5. American Psychological Association, "Frequently Asked Questions about Institutional Review Boards," http://www.apa.org/advocacy/research/defending-research/review-boards.aspx.
6. Harriet A. Washington, "Too Many Given No Right to Refuse in Medical Trials," *New Scientist,* January 18, 2012, https://www.newscientist.com/

article/mg21328480-200-too-many-given-no-right-to-refuse-in-medical-trials.

7. Richard Morse, "Grimes v. Kennedy Krieger Institute—Nontherapeutic Research with Children," *AMA Journal of Ethics* 5, no. 11 (November 2003): 383–85.
8. Harriet A. Washington, *Medical Apartheid,* 158–85.
9. *Erika Grimes v. Kennedy Krieger Institute,* 366 Md 29, 782 A 2d 807 (2001).
10. *The Works of Jeremy Bentham: Now First Collected under the Superintendence of His Executor, John Bowring,* vol. 7, chapter xvii (Ann Arbor: University of Michigan Library, 1838), 285.
11. Richard Rabin, "The Lead Industry and Lead Water Pipes 'a Modest Campaign,'" *American Journal of Public Health* 98, no. 9 (2008): 1,584–92. doi: 10.2105/AJPH.2007.113555.
12. Kevin Drum, "How Did Lead Get into Our Gasoline Anyway?," *Mother Jones,* January 7, 2013, http://www.motherjones.com/kevin-drum/2013/01/how-did-lead-get-our-gasoline-anyway/.
13. "Knock occurs when fuel is prematurely ignited in the engine's cylinder, which degrades efficiency and can be damaging to the engine. Knock is virtually unknown to modern drivers. This is primarily because fuels contain an oxygenate that prevents knock by adding oxygen to the fuel." Environmental and Energy Study Institute Fact Sheet, "A Brief History of Octane in Gasoline: From Lead to Ethanol," March 2016, http://www.ourenergypolicy.org/wp-content/uploads/2016/04/FactSheet_Octane_History_2016.pdf.
14. Kevin Drum, "Why Is Murder Down in São Paulo? The Answer is…" *Mother Jones,* August 2, 2013, https://www.motherjones.com/kevin-drum/2013/08/murder-sao-paulo-lead-ethanol/.
15. C. B. Grey and A. R. Varcoe, "Octane, Clean Air, and Renewable Fuels: A Modest Step Toward Energy Independence," *Texas Review of Law & Politics,* January 2006.
16. David Rosner and Gerald Markowitz, "It Goes Way Beyond Flint: America Has a Lead Crisis That Stretches from Coast to Coast," *In These Times,* February 9, 2016, http://inthesetimes.com/article/18835/americas-coast-to-coast-toxic-crisis.
17. Drum, "How Did Lead Get into Our Gasoline Anyway?"
18. Libby Kane, "How Much You Have to Earn to Be Considered Middle Class in Every US State," *Business Insider,* April 2, 2015, http://www.businessinsider.com/middle-class-in-every-us-state-2015-4; Paul Orum, Richard Moore, Michele Roberts, Joaquin Sanchez, *Who's in Danger: A Demographic Analysis of Chemical Disaster Vulnerability Zones,* https://comingcleaninc.org/assets/media/images/Reports/Who%27s%20in%20Danger%20Report%20FINAL.pdf.
19. Derrick Z. Jackson, "Environmental Justice? Unjust Coverage of the Flint Water Crisis," Joan Shorenstein Centre on Media Politics and Policy, July 11, 2017, https://shorensteincenter.org/environmental-justice-unjust-coverage-of-the-flint-water-crisis/.

20. Centers for Disease Control and Prevention, "Sources of Lead," May 29, 2015, https://www.cdc.gov/nceh/lead/tips/sources.htm.
21. Blanc-de-Neige, "The Characteristics and Uses of Zinc White," *The Chameleon,* January 1900, 23, https://books.google.com/books?id=z0lBAQAAMAAJ&pg=RA1-PA33&lpg=RA1-PA33&dq=The+Chameleon+zinc+white&source=bl&ots=O93FHFJIKW&sig=hBEtI0jIIXEAN3ykxlbcBw-UPRE&hl=en&sa=X&ved=0ahUKEwiygvLKjvrYAhVC2VMKHeDoDRQQ6AEIPTAI#v=onepage&q=The%20Chameleon%20zinc%20white&f=false.
22. Quoted in Lilly Fowler, "After Flicking Away Lawsuits, Lead Industry Goes for a Final Knockout," August 8, 2013, https://www.fairwarning.org/2013/08/after-flicking-away-lawsuits-lead-industry-goes-for-a-final-knockout/.
23. Ibid.
24. Rosner and Markowitz, "It Goes Way Beyond Flint."
25. "Odd Gas Kills One, Makes Four Insane," *New York Times,* October 27, 1924, 1.
26. Herbert L. Needleman, "History of Lead Poisoning in the World," the Center for Biological Diversity, https://www.biologicaldiversity.org/campaigns/get_the_lead_out/pdfs/health/Needleman_1999.pdf.
27. Jonah Lehrer, "The Crime of Lead Exposure," *Wired,* June 1, 2011, https://www.wired.com/2011/06/the-crime-of-lead-exposure/.
28. Kevin Drum, "Lead: America's Real Criminal Element—The Hidden Villain behind Violent Crime, Lower IQs, and even the ADHD Epidemic," *Mother Jones,* February 11, 2016, https://www.motherjones.com/environment/2016/02/lead-exposure-gasoline-crime-increase-children-health/#.
29. Richard Rabin, M.D., August 8, 2013, comment on Lilly Fowler, "After Flicking Away Lawsuits, Lead Industry Goes for a Final Knockout," https://www.fairwarning.org/2013/08/after-flicking-away-lawsuits-lead-industry-goes-for-a-final-knockout.
30. Letter to Mr. Felix E. Wormer, Assistant Secretary United States Department of the Interior, Washington, D.C., from Bowditch Lead Industries Association, July 11, 1956. Collection of the American Heritage Center, University of Wyoming, item N13644.
31. Lilly Fowler, "California Cities Seek $1 Billion Settlement for Lead Paint-Related Health Care Costs," Center for Public Integrity, August 8, 2013, https://www.publicintegrity.org/2013/08/08/13137/california-cities-seek-1-billion-settlement-lead-paint-related-health-care-costs.
32. David Michaels, *Doubt Is Their Product: How Industry's Assault on Science Threatens Your Health* (New York: Oxford University Press, 2008).
33. Harriet A. Washington, "Flacking for Big Pharma," *The American Scholar,* Summer 2011, 22–34, https://theamericanscholar.org/flacking-for-big-pharma/#.Wm18OUtOmB0.
34. A. L. Fairchild, D. Rosner, J. Colgrove, R. Bayer, and L. P. Fried, "The Exodus of Public Health: What History Can Tell Us about the Future,"

American Journal of Public Health 100(1) (2010): 54–63, doi:10.2105/AJPH.2009.163956.

35. Ibid., 54, 60.
36. Cited in Fowler, "California Cities Seek $1 Billion Settlement for Lead Paint-Related Health Care Costs."
37. Harriet A. Washington, *Deadly Monopolies: The Shocking Corporate Takeover of Life Itself—And the Consequences for Your Health and Our Medical Future* (New York: Doubleday, 2011), 270–76.
38. "Fight Poverty, Not Patents," urged Carl Bildt in a January 3, 2003, *Wall Street Journal* op-ed; "Poverty, Not Patents, Is to Blame," declared the headline of James Shikwati's June 7, 2004, article in *Business Day* (Johannesburg), www.allafrica.com; "Poverty, Not Pharmaceutical Patents, Leading Factor in Lack of Access to Medicine in Developing Countries," the press-release site PRWeb claimed in February 2009; "Poverty, not patents, imposes the greater limitation on access," summarizes a 2009 article in the *Journal of Health Affairs* whose title reads "Poverty, Not Pharmaceutical Patents, Leading Factor in Lack of Access to Medicine in Developing Countries." "Poverty and sickness won't be cured by fighting patents," opined Franklin Cudjoe in a widely reproduced January 2011 opinion piece, and in January 2011 Keith Martin, M.D., M.P., published an *Edmonton Sun* opinion piece, also widely reproduced, entitled, "Poverty, Not Patents," cited in Harriet A. Washington, *Deadly Monopolies,* 393.
39. *Robert L. Ziegfeld v. Lead Industries Association.*
40. Citied in Fowler, "California Cities Seek $1 Billion Settlement for Lead Paint-Related Health Care Costs."
41. Harold C. Passer, "Housing Bias Hits Hard at Educated Negro," *Rochester Times-Union,* June 19, 1960; box 4, folder 14 contains "Housing Bias Hits Hard at Educated Negro." Original. 2 pp. "The Negro Problem in America with Special Reference to Rochester." Summary of the "Brick Forum" series by Harold C. Passer. Concludes that a lot of tensions are due to housing and has five attached articles from the *Democrat, Chronicle,* and *Times-Union* on housing, ranging from 1952–60. TS. 8 pp. "City Is Caught in US Dilemma of Race Relations." Original. 2 pp. "Better Housing Is the Negroes' Strongest Cry." Original and PC of first page. 2 pp. "Housing Bias Hits Hard at Educated Negro." Original. 2 pp. "SCAD Becomes a Symbol to Negro Community." Original. 2 pp. "Negro Remains Man Apart in Community." Original. Two copies of second page. 3 pp. "No Man Should Feel His Color Bars His Rise." *Times-Union,* June 10, 1960. Original. 1 p. in "Dr. Walter Cooper Papers" collection, University of Rochester Rush Rhees Library, call number: d.385; dates: 1939–2011; physical description: 22 boxes.
42. Alexandra Stevenson and Matthew Goldstein, "Seller-Financed Deals Are Putting Poor People in Lead-Tainted Homes," *New York Times,* December 26, 2016, https://www.nytimes.com/2016/12/26/business/dealbook/seller-financed-home-sales-poor-people-lead-paint.html.

43. Erin Schumaker and Alissa Scheller, "Lead Poisoning Is Still a Public Health Crisis for African-Americans," *Huffington Post,* July 13, 2015, updated February 27, 2017, https://www.huffingtonpost.com/2015/07/13/black-children-at-risk-for-lead-poisoning-_n_7672920.html.
44. "Blood Lead Levels—United States, 1988–1991," *Morbidity and Mortality Weekly Report* 43, no. 30 (1994): 545–48, https://www.cdc.gov/mmwr/preview/mmwrhtml/00032080.htm#00000298.gif.
45. Ibid.
46. World Health Organization Press, "Childhood Lead Poisoning," 2010, http://www.who.int/ceh/publications/leadguidance.pdf.
47. H. L. Needleman et al., "The Long-Term Effects of Exposure to Low Doses of Lead in Childhood: An 11-Year Follow-up Report," *New England Journal of Medicine* 322, no. 2 (1990): 83–88; D. C. Bellinger, K. M. Stiles, and H. L. Needleman, "Low-Level Lead Exposure, Intelligence and Academic Achievement: A Long-Term Follow-up Study," Pediatrics 90, no. 6 (1992): 855–61; W. J. Rogan et al., "The Effect of Chelation Therapy with Succimer on Neuropsychological Development in Children Exposed to Lead," *New England Journal of Medicine* 344, no. 19 (2001): 1,421–26.
48. World Health Organization Press, "Childhood Lead Poisoning"; Needleman et al., "The Long-Term Effects of Exposure to Low Doses of Lead in Childhood"; Bellinger et al., "Low-Level Lead Exposure, Intelligence and Academic Achievement"; Rogan et al., "The Effect of Chelation Therapy with Succimer on Neuropsychological Development in Children Exposed to Lead."
49. M. R. Basha et al., "The Fetal Basis of Amyoidogenesis: Exposure to Lead and Latent Overexpression of Amyloid Precursor Protein and ß-Amyloid in the Aging Brain," *Journal of Neuroscience* 25: 823–29; W. J. Rogan, et al., "The Effect of Chelation Therapy with Succimer on Neuropsychological Development in Children Exposed to Lead," *New England Journal of Medicine* 344, no. 19 (2001): 1,421–26; J. Wu et al., "Alzheimers' Disease (AD) Like Pathology in Aged Monkeys after Infantile Exposure to Environmental Metal Lead (Pb): Evidence for a Developmental Origin and Environmental Link for AD," *Journal of Neuroscience* 28 (2008): 3–9; J. R. Pilsner et al., "Influence of Prenatal Lead Exposure on Genomic Methylation of Cord Blood DNA," *Environmental Health Perspectives* 117 (2009): 1,466–71; N. Pawlas, "Oxidative Damage of DNA in Subjects Occupationally Exposed to Lead," *Advances in Clinical and Experimental Medicine* 26, no. 6 (September 2017): 939–45, doi:10.17219/acem/64682, https://www.ncbi.nlm.nih.gov/pubmed/29068594; Cheng Guo et al., "Association between Oxidative DNA Damage and Risk of Colorectal Cancer: Sensitive Deter mination of Urinary 8-Hydroxy-2′-deoxyguanosine by UPLC-MS/MS," *Analysis Scientific Reports* 6 (2016): 32581, https://www.nature.com/articles/srep32581; R. R. Dietert and M. S. Piepenbrink, "Lead and Immune Function," *Critical Review of Toxicology* 36, no. 4 (April 2006): 359–85.
50. Ibid.

51. J. S. Schneider and E. DeCamp, "Postnatal Lead Poisoning Impairs Behavioral Recovery Following Brain Damage," *Neurotoxicology* 28, no. 6 (2007)): 1,153–57.
52. Bill Quigley, "The 'Shocking' Statistics of Racial Disparity in Baltimore: A Look at the City Through Numbers That Tell a Story," *Common Dreams,* April 28, 2015, https://www.commondreams.org/views/2015/04/28/shocking-statistics-racial-disparity-baltimore.
53. Deborah W. Denno, "Considering Lead Poisoning as a Criminal Defense," *Fordham Urban Law Journal* 20, no. 31992. The paper's abstract reads, "A brief survey of the causes and consequences of lead poisoning suggests that lead poisoning is pervasive, particularly among blacks in urban communities, that it's [*sic*] effects can be debilitating, and that it has been linked to disciplinary problems, aggression, and repetitive and oftentimes violent crime. There is a legitimate question, then, about whether lead poisoning should be considered a viable criminal defense. The question of whether lead poisoning should be a defense is perhaps most appropriately placed in the context of debates regarding free will, determinism, and the ability of social scientists to predict the course of any one individual's behavior."
54. Jessica Wolpaw Reyes, "Lead Exposure and Behavior: Effects on Antisocial and Risky Behavior among Children and Adolescents," *Economic Enquiry,* February 12, 2015, http://www.nber.org/papers/w20366.
55. Kevin Drum, "Lead: America's Real Criminal Element—The Hidden Villain Behind Violent Crime, Lower IQs, and Even the ADHD Epidemic," *Mother Jones,* February 11, 2016.
56. Quigley, "The 'Shocking' Statistics of Racial Disparity in Baltimore."
57. Mike Hellgren, "Criminal Cases Dropped after Video Allegedly Shows Officer Planting Drugs," *CBS News,* https://www.youtube.com/watch?v=3Q7qU6RwvP0.
58. Harriet A. Washington, "Base Assumptions? Racial Aspects of U.S. DNA Forensics," in *Genetic Suspects: Global Governance of Forensic DNA Profiling and Databasing,* eds. Richard Hindmarsh and Barbara Prainsack (Cambridge: Cambridge University Press, 2010), 66.
59. Peter Neufeld telephone communication with the author. Also, Washington, "Base Assumptions?," 66.
60. Adam K. Liptak, "Study Suspects Thousands of False Convictions," *New York Times,* April 19, 2004, 15.
61. B. Weiss, "Food Additives and Environmental Chemicals as Sources of Childhood Behavior Disorders," *Journal of the American Academy Child Psychiatry* 21 (1982): 144–52.
62. Harriet A. Washington, "Prudence and the Pill: Testing Thalidomide in the Global South," *The Biopolitical Times,* December 13, 2010, https://www.geneticsandsociety.org/biopolitical-times/prudence-and-pill-testing-thalidomide-global-south.
63. From David Rosner and Gerald Markowitz, "Standing up to the Lead Industry: An Interview with Herbert Needleman," *Public Health Reports*

120(3) (2005): 330–37, doi:10.1177/003335490512000319, https://www.ncbi.nlm.nih.gov/pmc/articles/PMC1497712/.

64. Herbert L. Needleman et al., "Deficits in Psychologic and Classroom Performance of Children with Elevated Dentine Lead Levels," *New England Journal of Medicine* 300 (1979): 689–95, doi:10.1056/NEJM197903293001301.
65. Herbert L. Needleman, "The Persistent Threat of Lead: A Singular Opportunity," *American Journal of Public Health* 79, no. 5 (May 1989): 643.
66. Rosner and Markowitz, *Lead Wars,* 135.
67. Rosner and Markowitz, "Standing up to the Lead Industry," 336.
68. CDRH Division of Industry Communication and Education, "FDA Warns against Using Magellan Diagnostics LeadCare Testing Systems with Blood Obtained from a Vein: FDA Safety Communication," U.S. Food and Drug Administration, May 17, 2017, https://www.fda.gov/MedicalDevices/Safety/AlertsandNotices/ucm558733.htm.
69. Timothy B. Wheeler and Luke Broadwater, "Lead Paint: Despite Progress, Hundreds of Maryland Children Still Poisoned," *Baltimore Sun,* December 5, 2015, http://www.baltimoresun.com/news/maryland/investigations/bs-md-lead-poisoning-gaps-20151213-story.htm.
70. Jon Swaine, Oliver Laughland, Jamiles Lartey, and Ciara McCarthy, "Young Black Men Killed by US Police at Highest Rate in Year of 1,134 Deaths," *The Guardian,* December 31, 2015, https://www.theguardian.com/us-news/2015/dec/31/the-counted-police-killings-2015-young-black-men.
71. M. B. Pell and Joshua Schneyer, "Reuters Finds Lead Levels Higher Than Flint's in Thousands of Locales," December 19, 2016, http://www.reuters.com/investigates/special-report/usa-lead-testing/.
72. Wheeler and Broadwater, "Lead Paint: Despite Progress, Hundreds of Maryland Children Still Poisoned."
73. Gerald Markowitz and David Rosner, *Deceit and Denial: The Deadly Politics of Industrial Pollution* (Berkeley: University of California Press; New York: Milbank Memorial Fund, 2002).
74. Michael Dresser and Timothy B. Wheeler, "Housing Chief Suggests Mothers May Deliberately Expose Children to Lead," *Baltimore Sun,* August 14, 2016, http://www.baltimoresun.com/news/maryland/bs-md-lead-liability-20150814-story.html; Josh Hicks, "A Look at Kenneth Holt, Maryland's Embattled Housing Secretary," *Washington Post,* September 1, 2015, https://www.washingtonpost.com/local/md-politics/a-look-at-kenneth-holt-marylands-embattled-housing-secretary/2015/09/01/96b85750-4848-11e5-8ab4-c73967a143d3_story.html?utm_term=.121da1dfc5b99.
75. Wheeler and Broadwater, "Lead Paint: Despite Progress, Hundreds of Maryland Children Still Poisoned."
76. Ibid.
77. Ron Fonger, "General Motors Shutting Off Flint River Water at Engine Plant over Corrosion Worries," Michigan Live, October 13, 2014, updated January 17, 2015, http://www.mlive.com/news/flint/index.ssf/2014/10/general_motors_wont_use_flint.html.

78. Maureen McDonald, "9,000 Flint Residents Sue over Tainted Water," *The Daily Beast,* November 29, 2017, https://www.thedailybeast.com/9000-flint-residents-sue-over-tainted-water; see also, Julie Bosman, Monica Davey, and Mitch Smith, "As Water Problems Grew, Officials Belittled Complaints from Flint," January 20, 2016, *New York Times,* https://www.nytimes.com/2016/01/21/us/flint-michigan-lead-water-crisis.html.
79. David Rosner and Gerald Markowitz, "It Goes Way Beyond Flint."
80. Mona Hanna-Attisha et al., "Elevated Blood Lead Levels in Children Associated with the Flint Drinking Water Crisis: A Spatial Analysis of Risk and Public Health Response," *American Journal of Public Health* 106, no. 2 (February 2016): 283–90.
81. David R. Williams, Naomi Priest, and Norman Anderson, "Understanding Associations between Race, Socioeconomic Status and Health: Patterns and Prospects," *Health Psychology* 35, no. 4 (April 2016): 407–11, doi:10.1037/hea0000242.
82. United States Census Bureau, "Community Facts," American Fact Finder, n.d., https://factfinder.census.gov/faces/nav/jsf/pages/index.xhtml.
83. U.S. Census Bureau, 2014.
84. Williams, Priest, and Anderson, "Understanding Associations between Race, Socioeconomic Status and Health."
85. Goodnough, "Flint Weighs Scope of Harm to Children Caused by Lead in Water."
86. Nanhua Zhang et al., "Early Childhood Lead Exposure and Academic Achievement: Evidence from Detroit Public Schools, 2008–2010," *American Journal of Public Health,* February 6, 2012, http://ajph.aphapublications.org/doi/10.2105/AJPH.2012.301164.
87. Lynne Peeples, "'Little Things Matter' Exposes Big Threat to Children's Brains," *Huffington Post,* November 20, 2014, updated December 6, 2017, https://www.huffingtonpost.com/2014/11/20/toxins-children-brain-little-things-matter_n_6189726.html4.
88. Rick Nevin, "How Lead Exposure Relates to Temporal Changes in IQ, Violent Crime, and Unwed Pregnancy," *Environmental Research* 83, no. 1 (May 2000): 1–22, https://www.sciencedirect.com/science/article/pii/S0013935199940458.
89. Ibid.
90. Cited in Peeples, "'Little Things Matter.'"
91. Narjas Zatat, "Pesticides Can Cause Brain Damage and Organic Food Is the Future, EU Report Says," *The Independent,* June 3, 2017, 31, https://www.independent.co.uk/environment/environment-report-eu-pesticide-link-brain-damage-lower-iq-billions-of-pounds-lost-organic-food-a7771056.html.
92. Chad Livengood, "Mayor: Water Fix Could Cost as Much as 1.5 Billion," *Detroit News,* January 7, 2016, http://www.detroitnews.com/story/news/politics/2016/01/07/flint-water/78404218/.

93. Goodnough, "Flint Weighs Scope of Harm to Children Caused by Lead in Water."
94. Rosner and Markowitz, "It Goes Way Beyond Flint."
95. Justin Gardner, "Navajo Water Supply Is More Horrific Than Flint, but No One Cares Because They're Native American," The Free Thought Project, January 31, 2016, https://www.google.com/search?q=http%3A%2F%2Fthefreethoughtproject.com%2Fnavajo-water-supply-horrific&ie=utf-8&oe=utf-8.
96. Laura Orlando, "Why Your Water Could Be Worse Than Flint's," *In These Times,* March 28, 2016, https://www.google.com/search?q=http%3A%2F%2Finthesetimes.com%2Farticle%2F18951%2Fis-your-water-worse-than-f…&ie=utf-8&oe=utf-8.
97. Mireya Navarro and William K. Rashbaum, "U.S. Investigating Elevated Blood Lead Levels in New York's Public Housing," *New York Times,* March 16, 2016, https://www.nytimes.com/2016/03/17/nyregion/us-investigating-elevated-blood-lead-levels-in-new-yorks-public-housing.html.
98. Alison Young and Peter Eisler, "Some Neighborhoods Dangerously Contaminated by Lead Fallout," *USA Today,* March 20, 2012, http://usatoday30.usatoday.com/news/nation/story/2012-04-20/smelting-lead-contamination-soil-testing/54420418/.
99. Ibid.

Chapter 3: Poisoned World: The Racial Gradient of Environmental Neurotoxins

1. Metropolitan Transportation Authority, "MTA Celebrates Re-opening of Mother Clara Hale Depot with Ribbon-Cutting," November 20, 2014, http://www.mta.info/news-nyct-buses-nyct-bus/2014/11/20/mta-celebrates-re-opening-mother-clara-hale-depot-ribbon-cutting.
2. Robert D. Bullard, "Environmental Justice for All," *Issues & Views,* http://www.uky.edu/~tmute2/GEI-Web/password-protect/GEI-readings/Bullard-Environmental%20justice%20for%20all.pdf.
3. Robert Bullard, "African Americans Need a Strong and Independent Federal EPA," *OpEdNews,* February 21, 2017.
4. B. P. Lanphear, "The Impact of Toxins on the Developing Brain," *Annual Review of Public Health* 36 (2015): 211–30.
5. The Environmental Justice Climate Change Initiative, "Air of Injustice: African Americans and Power Plant Pollution," October 2002, https://www.scribd.com/document/33842120/Air-of-Injustice-African-Americans-and-Power-Plant-Pollution.
6. Bullard, "Environmental Justice for All."
7. Ibid.
8. Luke Cole and Sheila R. Foster, *From the Ground Up: Environmental Racism and the Rise of the Environmental Justice Movement* (New York: New York University Press, 2001); Errol Schweizer, "Environmental Justice: An Interview with Robert Bullard," *Earth First! Journal,* July 1999.

9. Bullard, "Environmental Justice for All."
10. *Bean v. Southwestern Waste Management Corporation—Significance, Waste Management in Houston, Laches and State Action, Impact, Further Readings,* December 12, 1979, http://law.jrank.org/pages/13187/Bean-v-Southwestern-Waste-Management-Corp.html.
11. Ibid.
12. In response to the threat of hazardous waste sites, typified by the 1978 identification of a hazardous seventy-acre landfill in the Love Canal neighborhood of Niagara Falls, the federal government designed a program to fund the cleanup of especially harmful pollutants. Sites managed under this program are referred to as Superfund sites.
13. Asthma and Allergy Foundation of America, "Ethnic Disparities in Asthma," http://www.aafa.org/page/burden-of-asthma-on-minorities.aspx; see also, Office of Disease Prevention and Health Promotion, "Disparities," http://www.healthypeople.gov/2020/about/foundation-health-measures/Disparities.
14. Katherine Kornei, "Here Are Some of the World's Worst Cities for Air Quality," *Science,* March 21, 2017, http://www.sciencemag.org/news/2017/03/here-are-some-world-s-worst-cities-air-quality.
15. Justin Worland, "Preterm Births Linked to Air Pollution Cost Billions in the U.S.," *Time,* March 28, 2016.
16. "The *Lancet* Commission on Pollution and Health," *The Lancet* 391, October 19, 2017, https://www.thelancet.com/journals/lancet/article/PIIS0140-6736(17)32345-0/fulltext.
17. The NAACP, the Clean Air Task Force, and the National Medical Association, "Fumes across the Fence-Line: The Health Impacts of Air Pollution from Oil and Gas Facilities on African American Communities," November 14, 2017, https://www.naacp.org/latest/fumes-across-fence-line-new-study-naacp-clean-air-task-force-national-medical-association/.
18. Donna Owens, "African-Americans Face More Pollution-Related Health Hazards, New Report Shows," NBC News, November 14, 2017.
19. Ibid.
20. L. J. Seidman et al., "The Relationship of Prenatal and Perinatal Complications to Cognitive Functioning at Age 7 in the New England Cohorts of the National Collaborative Perinatal Project," *Schizophrenia Bulletin* 26, no. 2 (200): 309–21.
21. Michael Anastario et al., "Impact of Fetal Versus Oerinatal Hypoxia on Sex Differences in Childhood Outcomes: Developmental Timing Matters," *Social Psychiatry and Psychiatric Epidemiology* 47 (2017): 455–64.
22. Open Access Government, "An Emerging Environmental Health Concern: Impacts of Air Pollution on the Brain," January 18, 2018, https://www.openaccessgovernment.org/emerging-environmental-health-concern-impacts-air-pollution-brain/41301/.
23. Justin Worland, "Toxic Air Pollution Can Penetrate the Brain: Study," *Time,* September 6, 2016.

24. S. Franco Suglia et al., "Association of Black Carbon with Cognition among Children in a Prospective Birth Cohort Study," *American Journal of Epidemiology* 167, no. 3 (September 21, 2008): 280–86.
25. "The evidence to date is coherent in that exposure to a range of largely traffic-related pollutants has been associated with quantifiable impairment of brain development in the young and cognitive decline in the elderly." Angela Clifford et al., "Exposure to Air Pollution and Cognitive Functioning across the Life Course—A Systematic Literature Review," *Environmental Research* 147 (May 2016): 383–98.
26. "Including the Kaufman Brief Intelligence Test (K-BIT) and the Wide Range Assessment of Memory and Learning (WRAML). The K-BIT is an individually administered test of verbal and nonverbal intelligence (15). Two subscales, vocabulary and matrices, comprise the test, as well as a composite intelligence quotient (IQ) score. The K-BIT has acceptable correlation with the widely used Wechsler verbal performance and full-scale IQ scores (16); validation studies have been conducted for children less than 7 years of age with normative data available (17). The WRAML is a well-standardized psychometric instrument that allows evaluation of a child's ability to actively learn and memorize a variety of information (18, 19). The WRAML includes subscales on verbal memory, visual memory, and learning and an overall general index scale. It has been normed for children aged 5–17 years among racially diverse groups, including minorities. All measures are expressed as standardized scores, which represent the score of the individual taking the test relative to scores obtained by children of the same age and gender in the standardization sample. All scores have a mean of 100 and a standard deviation of 15." Angela Clifford et al., "Exposure to Air Pollution and Cognitive Functioning across the Life Course—A Systematic Literature Review," *Environmental Research* 147 (May 2016): 385.
27. Ranft et al., "Long-term exposure to traffic-related particulate matter impairs cognitive function in the elderly," *Environmental Research* 109 (2009): 1,004–11, doi:10.1016/j.envres.2009.08.003.
28. According to a K-BIT scale, per Suglia et al., "Association of Black Carbon with Cognition among Children in a Prospective Birth Cohort Study."
29. "Pollution-Born Magnetites Found in the Brain Could Be Linked to Alzheimer [*sic*]," *Science and Technology Research News*, September 12, 2016, https://www.scienceandtechnologyresearchnews.com/pollution-born-magnetites-found-brain-linked-alzheimer/.
30. Soong Ho Kim et al., "Rapid doubling of Alzheimer's amyloid-β40 and 42 levels in brains of mice exposed to a nickel nanoparticle model of air pollution", *F1000Research* 1 (December 2012): 16, doi: 10.12688/f1000research.1-70.
31. Annie Sneed, "DDT, Other Environmental Toxins Linked to Late-Onset Alzheimer's Disease," *Scientific American*, February 10, 2014, https://www.scientificamerican.com/article/studies-link-ddt-oth.

32. David Shukman, "Pollution Particles 'Get into Brain,'" BBC News, September 5, 2016; see also, "Air Pollution: This Toxic Pollutant Could Harm the Brain," *Time,* http://time.com/4480016/air-pollution-health-effects/; Justin Worland, "Preterm Births Linked to Air Pollution Cost Billions in the U.S.," *Time,* http://time.com/4274355/air-pollution-preterm-birth/.
33. Using mice that are genetically altered to show the development of Down syndrome and Alzheimer's disease changes in the brain at six months, allowing researchers to seek prevention strategies for this pathophysiology, researchers have found that maternal choline supplementation protects against basal forebrain cholinergic neuron degeneration seen in these animals. B. J. Strupp et al., "Maternal Choline Supplementation: A Potential Prenatal Treatment for Down Syndrome and Alzheimer's Disease," *Current Alzheimer's Research* 13, no. 1 (2016): 97–106.
34. Justin Worland, "Toxic Air Pollution Can Penetrate the Brain: Study," *Time,* September 6, 2016.
35. G. Oberdörster et al., "Translocation of inhaled ultrafine particles to the brain," *Toxicology* 16 (2004): 437–45; A. Elder et al., "Translocation of Inhaled Ultrafine Manganese Oxide Particles to the Central Nervous System," *Environmental Health Perspectives* 114 (2006): 1,172–78.
36. J. L. Kirschvink, A., Kobayashi-Kirschvink, and B. J. Woodford, *Proceedings of the National Academy of Science USA* 89, no. 16 (1992): 7,683–87.
37. Shukman, "Pollution Particles 'Get into Brain.'"
38. Annie Sneed, "DDT, Other Environmental Toxins Linked to Late-Onset Alzheimer's Disease," *Scientific American,* February 10, 2014, https://www.scientificamerican.com/article/studies-link-ddt-other-environmental-toxins-to-late-onset-alzheimers-disease/.
39. Lilian Calderón-Garcidueñas et al., "Air Pollution, Cognitive Deficits and Brain Abnormalities: A Pilot Study with Children and Dogs," *Brain and Cognition* 68 (2008): 117–27, https://www.ncbi.nlm.nih.gov/pubmed/18550243.
40. U. Ranft et al., "Long-Term Exposure to Traffic-Related Particulate Matter Impairs Cognitive Function in the Elderly"; see also, "Residential Proximity to Nearest Major Roadway and Cognitive Function in Community-Dwelling Seniors: Results from the MOBILIZE Boston Study," *Journal of American Geriatrics Society* 60, no. 11 (November 1, 2012), https://www.ncbi.nlm.nih.gov/pubmed/23126566.
41. The NAACP, the Clean Air Task Force, and the National Medical Association, "Fumes Across the Fence-Line."
42. Ibid.
43. Robert Bullard, telephone interview with the author, July 22, 2017.
44. The CDC indicates that benzene causes short-term drowsiness, confusion, convulsions, and long-term blood problems including leukemia and impaired reproduction. According to the authors of an article published in the journal *Neurology,* a 2014 study of 2,143 utility workers found that "high exposure to solvents was significantly associated with poor cognition; for example, those highly exposed to chlorinated solvents

were at risk of impairment on the Mini-Mental State Examination (risk ratio 1.18; 95% confidence interval 1.06, 1.31), the Digit Symbol Substitution Test (1.54; 1.31, 1.82), semantic fluency test (1.33; 1.14, 1.55), and the Trail Making Test B (1.49; 1.25, 1.77)." Erica L. Sabbath et al., "Time May Not Fully Attenuate Solvent-Associated Cognitive Deficits in Highly Exposed Workers," *Neurology* 82 (2014): 1,716–23, https://www.ncbi.nlm.nih.gov/pmc/articles/PMC4032208/.

45. Haji Bahadar, Sara Mostafalou, and Mohammad Abdollahi, "Current Understandings and Perspectives on Non-Cancer Health Effects of Benzene: A Global Concern," *Toxicology and Applied Pharmacology* 276, no. 2 (April 15, 2014): 83–94, https://www.sciencedirect.com/science/article/pii/S0041008X1400057X.
46. Sabbath et al., "Time May Not Fully Attenuate Solvent-Associated Cognitive Deficits in Highly Exposed Workers."
47. Roberta Lo Pumo, "Long-Lasting Neurotoxicity of Prenatal Benzene Acute Exposure in Rats," *Toxicology* 223, no. 3 (June 15, 2006): 227–34, https://www.ncbi.nlm.nih.gov/pubmed/16698163; A. Lertxundi et al., "Exposure to Fine Particle Matter, Nitrogen Dioxide and Benzene During Pregnancy and Cognitive and Psychomotor Developments in Children at 15 Months of Age," *Environment International* 80 (2015): 33–40, https://www.ncbi.nlm.nih.gov/pubmed/25881275.
48. U.S. Department of Justice, Office of Public Affairs, "Tonawanda Coke to Pay $12 Million in Civil Penalties, Facility Improvements and Environmental Projects to Benefit Tonawanda Community," May 11, 2015, https://www.justice.gov/opa/pr/tonawanda-coke-pay-12-million-civil-penalties-facility-improvements-and-environmental.
49. The NAACP, the Clean Air Task Force, and the National Medical Association, "Fumes Across the Fence-Line."
50. Ibid.
51. P. Butterworth et al., "The Psychosocial Quality of Work Determines Whether Employment Has Benefits for Mental Health: Results from a Longitudinal National Household Panel Survey," *Occupational and Environmental Medicine* 68 (March 14, 2011): 806–12, http://oem.bmj.com/content/early/2011/02/26/oem.2010.059030.
52. Cynthia Shahan, "Major Poll Shows African Americans Support Clean Energy, Clean Jobs, and Clean Power Plan," *CleanTechnica*, November 23, 2015, https://cleantechnica.com/2015/11/23/major-poll-shows-african-americans-believe-shifting-clean-energy-will-create-jobs/.
53. Ibid.
54. Owens, "African-Americans Face More Pollution-Related Health Hazards, New Report Shows."
55. Gerald Markowitz and David Rosner, "'Unleashed on an Unsuspecting World': The Asbestos Information Association and Its Role in Perpetuating a National Epidemic," *American Journal of Public Health* 106, no. 5 (May 2016): 834–40, https://www.unboundmedicine.com/ccdm/view/Public-Health-News/11332/0/%22Unleashed_on_an_Unsuspecting

_World%22:_The_Asbestos_Information_Association_and_Its_Role_in_Perpetuating_a_National_Epidemic?ti=21.

56. Michael Asher, "Genetically Modified Organisms No Answer to Food Shortage," *The Star* (Kenya), October 4, 2012, https://www.the-star.co.ke/news/2012/10/04/gmos-no-answer-to-food-shortage_c686514.
57. Anne Outwater, "Monsanto and Syngenta—Funding Agra?," *Tanzania Daily News*, June 3, 2012, http://allafrica.com/stories/201206041269.html.
58. Environmental Working Group, "Poisoned by PCBs: A Lack of Control," *Chemical Industry Archives*, December 10, 2015, http://www.chemicalindustryarchives.org/dirtysecrets/anniston/2.asp; retrieved October 31, 2015.
59. Randall Fitzgerald, *The Hundred-Year Lie: How Food and Medicine Are Destroying Your Health* (New York: Plume, 2006).
60. Ted Schettler et al., "The DDT Question," *The Lancet* 356, no. 9236 (September 20, 2000): 1,189–90, http://www.thelancet.com/journals/lancet/article/PIIS0140-6,736(05)72883-X/fulltext.
61. V. Karri, M. Schuhmacher, and V. Kumar, "Heavy Metals (Pb, Cd, As and MeHg) as Risk Factors for Cognitive Dysfunction: A General Review of Metal Mixture Mechanism in Brain," *Environmental Toxicology and Pharmacology* 48 (September 26, 2016): 203–13, https://www.ncbi.nlm.nih.gov/pubmed/27816841 DOI: 10.1016/j.etap.2016.09.016.
62. *Owens v. Monsanto*, 96-CV-440, (N.D. Ala.), http://www.chemicalindustryarchives.org/dirtysecrets/annistonindepth/intro.asp.
63. Michael Grunwald, "Monsanto Hid Decades of Pollution," *Washington Post*, January 1, 2002, https://www.commondreams.org/headlines02/0101-02.htm.
64. Patricia Born, "City of Fort Myers Dumped Toxic Sludge in Dunbar," *News-Press*, June 12, 2017, https://www.news-press.com/story/news/2017/06/12/city-fort-myers-dumped-toxic-sludge-dunbar/369451001/.
65. Molly Tolins and Philip Landrigan, "The Developmental Neurotoxicity of Arsenic: Cognitive and Behavioral Consequences of Early Life Exposure," *Annals of Global Health* 80, no. 4 (July–August 2014): 303–14, https://doi.org/10.1016/j.aogh.2014.09.005.
66. "Former Toxic Dump Site Sits Near Dunbar Homes," *WINK News*, June 12, 2017, http://www.winknews.com/2017/06/12/former-toxic-dump-site-sits-empty-near-dunbar-homes/.
67. Ibid.
68. David R. Williams, "How Racism Makes Us Sick," TEDMED 2016, https://www.ted.com/talks/david_r_williams_how_racism_makes_us_sick/up-next.
69. Patricia Borns, "Is Fort Myers' Decision to Dump Toxic Sludge in Dunbar Racist?," *News Press*, December 13, 2017, https://www.news-press.com/story/news/2017/12/13/fort-myers-toxic-sludge-dump-racist/940962001/.
70. Andrea Melendez, "Steve Hilfiker, CEO of Environmental Risk Management, Gives a Statement about Work He Was Asked to Do for the City of Fort

Myers at the Home-a-rama Site Back in 2002," June 23, 2017, http://www.news-press.com/videos/news/2017/06/23/home-rama-steve-hilfiker/103137814/.

71. Ibid.
72. Ben Henry, "Toxic Dump Site in Dunbar Neighborhood Gains Statewide Attention," NBC News, June 16, 2017, http://www.nbc2.com/story/35684865/toxic-dump-site-in-dunb.
73. Molly Tolins, Mathuros Ruchirawat, and Philip Landrigan, "The Developmental Neurotoxicity of Arsenic: Cognitive and Behavioral Consequences of Early Life Exposure," *Annals of Global Health* 80, no. 4 (July–August 2014): 303–14, https://doi.org/10.1016/j.aogh.2014.09.005.
74. "Arsenic," Agency for Toxic Substances and Disease Registry Toxin, https://www.atsdr.cdc.gov/sites/toxzine/arsenic_toxzine.html.
75. The report listed these tests among those administered: Behavior Assessment System (BASC) for Children; Behavior Rating Inventory of Executive Functions (BRIEF); BSID-II, Bayley Scales of Infant Development-II; CADS-IV, Conners' ADHD DMS-IV Scales; CAT, Cognitive Abilities Test; Child Behavior Checklist (CBCL); Children's Category Test-Level II (CCT); CDI, Children's Depression Inventory; CELF-3, Clinical Evaluation of Language Fundamentals-Third Edition; COWAT, Controlled Oral Word Association Test; CPRS, Conners' Parent Rating Scale; CPRS-R, Conners' Parent Rating Scale—Revised; CPT, Continuous Performance Test; CRT-RC, Combined Raven's Test—The Rural in China Method; CTRS, Conners' Teachers Rating Scale; CTRS-R, Conners' Teachers Rating Scale—Revised; California Verbal Learning Test-Children (CVLT-c); Diagnostic and Statistical Manual of Mental Disorders (DSM-IV); EXIT25, Exit Interview; MAT, Math Achievement Test; MMSE, Mini Mental Status Examination; NLST, Number and Letter Sequencing Test; PM, Pattern Memory; PPVT, Peabody Picture Vocabulary Test; PSTs, Problem Solving Tests; RBANS, Repeatable Battery for the Assessment of Neuropsychological Status; RCPMT, Raven Colored Progressive Matrices Test; SA, Switching Attention; SD, Symbol Digit; Trails Making Test (TMTA/TMTB); urinary arsenic (UAs); VSAFD, Visual-Spatial Abilities with Figure Design; WASI, Wechsler Abbreviated Intelligence Scale; WBRS Wolke's Behavior Rating Scale; WISC, Wechsler Intelligence Scale for Children; WISC-III, Wechsler Intelligence Scale for Children-Third Edition; WISC-IV, Wechsler Intelligence Scale for Children-Fourth Edition; WISC-RM, Wechsler Intelligence Scale for Children-Revised Mexican Edition; WPPSI, Wechsler Pre-school and Primary Scale of Intelligence; WRAML, Wide Range Assessment of Memory and Learning; Wide Range Assessment of Visual Motor Ability (WRAVMA).
76. This is according to the 2000 U.S. census, although some news media make reference to a 75 percent African American demographic. "2016 U.S. Gazetteer Files," https://www2.census.gov/geo/docs/maps-data/data/gazetteer/2016_Gazetteer/2016_gaz_place_01.txt.

77. University of Michigan Triana Justice Page, "Environmental Justice Case Study: DDT Contamination: Triana, AL," http://www.umich.edu/~snre492/triana.html.
78. Mike Hollis, "The Persistence of a Poison," *Washington Post,* June 15, 1980, A2, https://www.washingtonpost.com/archive/politics/1980/06/15/the-persistence-of-a-poison/52ff7202-c586-47ee-b306-b6b20c6b9dd1/?utm_term=.f66722abb5a2.
79. Ibid.
80. Tom Tiede, "Town Grimly Waits Out DDT Scare," *The Paris (TX) News,* November 28, 1979, 16, https://www.newspapers.com/newspage/9504912/.
81. Hollis, "The Persistence of a Poison."
82. "DDT (dichlorodiphenyltrichloroethane)," Cornell University Extension Toxicology Network, http://pmep.cce.cornell.edu/profiles/extoxnet/carbaryl-dicrotophos/ddt-ext.html.
83. Luisa Torres-Sánchez et al., "Prenatal p,p´-DDE Exposure and Neurodevelopment among Children 3.5–5 Years of Age," *Environmental Health Perspectives* 121, no. 2 (February 2013): 263–68, https://www.ncbi.nlm.nih.gov/pubmed/23151722.
84. Philippe Grandjean, "News: Chlorpyrifos Drains Brains," *Chemical Brain Drain,* http://braindrain.dk/2018/11/chlorpyrifos-drains-brains/.
85. M. F. Bouchard et al., "Prenatal Exposure to Organophosphate Pesticides and IQ in 7-Year-Old Children," *Environmental Health Perspectives* 119, no. 8 (August 2011): 1,189–95, doi:10.1289/ehp.1003185; M. F. Bouchard et al., "Seven-Year Neurodevelopmental Scores and Prenatal Exposure to Chlorpyrifos, a Common Agricultural Pesticide," *Environmental Health Perspectives* 119, no. 8, 1,196–1,201, doi:10.1289/ehp.1003160; Bradley S. Peterson, "Effects of Prenatal Exposure to Air Pollutants (Polycyclic Aromatic Hydrocarbons) on the Development of Brain White Matter, Cognition, and Behavior in Later Childhood," *AMA Psychiatry* 72, no. 6 (2015): 531–40, doi:10.1001/jamapsychiatry.2015.57.
86. R. F. White et al., "Adult Neuropsychological Performance Following Prenatal and Early Postnatal Exposure to Tetrachloroethylene (PCE)-Contaminated Drinking Water," *Neurotoxicology and Teratology* 34, no. 3 (May–June 2012): 350–59, doi:10.1016/j.ntt.2012.04.001.
87. Philippe Grandjean, *Only One Chance: How Environmental Pollution Impairs Brain Development—And How to Protect the Brains of the Next Generation* (New York: Oxford University Press, 2015), 108.
88. Elise Roze et al.,"Children's Health Prenatal Exposure to Organohalogens, Including Brominated Flame Retardants, Influences Motor, Cognitive, and Behavioral Performance at School Age," *Environmental Health Perspectives* 117, no. 12 (2009): 1,953–58, https://www.ncbi.nlm.nih.gov/pmc/articles/PMC2799472/; doi:10.1289/ehp.0901015.
89. Philippe Grandjean, "Known Chemical Brain Drainers," http://braindrain.dk/known-chemical-brain-drainers/.

90. Oliver Millman, “Dakota Access Pipeline Company and Donald Trump Have Close Financial Ties,” *The Guardian*, October 26, 2016, https://www.theguardian.com/us-news/2016/oct/26/donald-trump-dakota-access-pipeline-investment-energy-transfer-partners.
91. Raul Garcia, “We’re Missing 90 Percent of the Dakota Access Pipeline Story,” Earthjustice, November 22, 2016, https://earthjustice.org/blog/2016-november/we-re-missing-90-percent-of-the-dakota-access-pipeline-story.
92. Brian Hauss, “Standing Rock Protest Groups Sued by Dakota Access Pipeline Company,” December 6, 2017, https://www.aclu.org/blog/free-speech/rights-protesters/standing.
93. “Dakota Access Has Already Leaked Oil and It’s Not Yet Even Operational,” *Democracy Now,* May 12, 2017, http://www.iflscience.com/environment/dakota-access-pipeline-has-been-leaking-oil-and-its-not-even-operational-yet/.
94. “Reservations about Toxic Waste: Native American Tribes Encouraged to Turn Down Lucrative Hazardous Disposal Deals,” *Scientific American,* n.d., https://www.scientificamerican.com/article/earth-talk-reservations-about-toxic-waste/.
95. World Health Organization, “Inorganic Mercury,” *Environmental Health Criteria* 118 (1991), http://apps.who.int/iris/handle/10665/40626.
96. P. B. Tchounwou, W. K. Ayensu, N. Ninashvili, and D. Sutton, “Environmental Exposure to Mercury and Its Toxicopathologic Implications for Public Health,” *Environmental Toxicology* 18, no. 3 (May 6, 2003): 149–75, https://www.ncbi.nlm.nih.gov/pubmed/12740802.
97. Jianping Xue et al., “Methyl Mercury Exposure from Fish Consumption in Vulnerable Racial/Ethnic Populations: Probabilistic SHEDS-Dietary Model Analyses Using 1999–2006 NHANES and 1990–2002 TDS Data,” *Science of the Total Environment* 414 (2012): 373–79.
98. Ralph Maughan, “50 Dirtiest Power Plants Named,” *The Wildlife News,* July 29, 2007, http://www.thewildlifenews.com/2007/07/29/50-dirtiest-us-power-plants-named/.
99. David O. Williams, “Navajo Generating Station Blamed for Haze over Grand Canyon, Respiratory Illnesses,” *Colorado Independent,* September 16, 2011, http://www.coloradoindependent.com/99627/navajo-generating-station-blamed-for-haze-over-grand-canyon-respiratory-illnesses.
100. Jim McGowan, “Desert Rock Energy Project,” Sithe Global, http://www.sitheglobal.com/projects/desertrock.cfm.
101. Judith Nies, “The Black Mesa Syndrome: Indian Lands, Black Gold,” *Orion Magazine,* Summer 1998, https://orionmagazine.org/article/the-black-mesa-syndrome/.
102. Susan Gallagher, “Proposal Would Move Mining onto Crow Reservation,” Associated Press, April 5, 2008, http://bismarcktribune.com/news/state-and-regional/proposal-would-move-mining-onto-reservation/article_53c5af7e-92dc-597d-adac-37b09571e75e.html.

103. Sally Hardin and Angelica Lujan, "Trump's EPA Poised to Undo Progress on Mercury Pollution Reduction," Center for American Progress, December 18, 2018, https://www.americanprogress.org/issues/green/reports/2018/12/18/464269/trumps-epa-poised-undo-progress-mercury-pollution-reduction/.
104. Juliet Eilperin and Brady Dennis, "In rollback of mercury rule, Trump could revamp how government values human health," *Washington Post,* October 1, 2018, available at https://www.washingtonpost.com/energy-environment/2018/10/01/rollback-mercury-rule-trump-could-revamp-how-government-values-human-health/?utm_term=.ef43fe3f24a0.
105. Clair Johnson, "Northern Cheyenne Coal Future 'Delicate Issue,'" *Billings Gazette,* November 16, 2006, http://web.archive.org/web/20080902091633/http://billingsgazette.net/articles/2006/11/16/news/local/45-coal.txt; Maughan, "50 Dirtiest Power Plants Named."
106. Johnson, "Northern Cheyenne Coal Future 'Delicate Issue.'"
107. Maughan, "50 Dirtiest Power Plants Named."
108. "The aggregate loss in cognition associated with MeHg [methylmercury] exposure in the 2000 U.S. birth cohort was estimated using two previously published dose-response models that relate increases in cord blood Hg concentrations with decrements in IQ. MeHg exposure was assumed not to be correlated with native cognitive ability. Previously published estimates were used to estimate economic costs of MR caused by MeHg." L. Trasande et al., "Mental Retardation and Prenatal Methylmercury Toxicity," *American Journal of Industrial Medicine* 49, no. 3 (March 2006): 153–58.
109. Ibid.

Chapter 4: Prenatal Policies: Protecting the Developing Brain

1. Ellen Griffith Spears, *Baptized in PCBs: Race, Pollution, and Justice in an All-American Town* (Chapel Hill: University of North Carolina Press, 2006).
2. Tom Neltner, "When It Comes to Lead, Formula-Fed Infants Get Most from Water and Toddlers from Food, but for Highest Exposed Children the Main Source of Lead is Soil and Dust," Environmental Defense Fund, February 3, 2017, http://blogs.edf.org/health/2017/02/03/sources-of-lead-exposure-infant-toddler/.
3. Ibid.
4. Kaiser Health News, "Lead in 20 Percent of Baby Food Samples," Black Health Matters.com, http://www.blackhealthmatters.com/nutrition-fitness/nutrition/food-news/lead-in-20-percent-of-baby-food-samples/.
5. Neltner, "When It Comes to Lead, Formula-Fed Infants Get Most from Water and Toddlers from Food, but for Highest Exposed Children the Main Source of Lead is Soil and Dust."
6. Environmental Working Group, "2019 Shopper's Guide to Pesticides in Produce," March 20, 2019, https://www.ewg.org/foodnews/summary.php; see also Liza Gross, "Over Half of Samples of Kale Tainted with Possible Cancer-Causing Chemical," *The Nation,* March 20, 2019;

https://www.thenation.com/article/pesticides-farmworkers-agriculture/.

7. Ruth A. Lawrence, *Breastfeeding: A Guide for the Medical Profession,* 8th ed. (New York: Elsevier, 2015).
8. Ibid.
9. Geoff Der, G. David Batty, and Ian Deary, "Effect of Breastfeeding on Intelligence in Children: Prospective Study, Sibling Pairs Analysis, and Meta-Analysis," *British Medical Journal* 333, no. 7575 (2006); R. Masters, "Brain Biochemistry and Social Status: The Neurotoxicity Hypothesis," in *Intelligence, Political Inequality, and Public Policy,* ed. Elliot White (New York: Praeger, 1997), 141–83.
10. Salomi Kafouri et al., "Breastfeeding and Brain Structure in Adolescence," *International Journal of Epidemiology* 42 (2013): 150–59, https://academic.oup.com/ije/article/42/1/150/695203 doi:10.1093/ije/dys172.
11. Although both the medical and popular literature tend to refer solely to "mothers" when discussing prenatal health, I prefer the less biased "parents," which acknowledges the key role and major contributions of fathers of color. Although about one in three American fathers lived apart from their children in 2009, some men of color are the primary or sole caregivers for their children. In the case of African American fathers, their semantic exclusion is based on the myth that they tend to be absent parents, but recent studies show that they tend to be more involved in their children's lives than fathers of other ethnicities. When appropriate, I include fathers by referring to "parents." Jo Jones and William Mosher, "Fathers' Involvement with Their Children: United States, 2006–2010," Division of Vital Statistics National Health Statistics Report Number 71n, December 20, 2013, https://pdfs.semanticscholar.org/9140/a75bcccc5b4c00a0f1b077a56597130ae983.pdf. See also Howard Dubowitz et al., "Low-Income African American Fathers' Involvement in Children's Lives: Implications for Practitioners," *Journal of Family Social Work* 10, no. 1 (2006): 25–41, http://www.tandfonline.com/doi/abs/10.1300/J039v10n01_02?journalCode=wfsw20.
12. Kafouri et al., "Breastfeeding and Brain Structure in Adolescence."
13. Girma Berhanu, "Academic Racism: Lynn's and Kanazawa's Ill-Considered Theory of Racial Differences in Intelligence," *Education Review* 14, no. 12 (2011): 3–4, https://pdfs.semanticscholar.org/4306/7f2587aee7c9def0a84b24120d3bbd00bd2b.pdf.
14. Jonah Lehrer, "Inside the Baby Mind," *Boston Globe,* April 26, 2009, http://archive.boston.com/bostonglobe/ideas/articles/2009/04/26/inside_the_baby_mind/?page=1.
15. Barbara Ehrenreich, *Nickel and Dimed: On (Not) Getting by in America* (New York: Picador, 2011), 27.
16. H. Rindermann and S. Pichelmann, "Future Cognitive Ability: US IQ Prediction until 2060 Based on NAEP," *PLoS ONE* 10(10) (2015): e0138412, doi:10.1371/journal.pone.0138412, https://www.ncbi.nlm.nih.gov/pmc/articles/PMC4603674/pdf/pone.0138412.pdf.

17. E. A. Hanushek, P. E. Peterson, and L. Woessmann, *Endangering Prosperity* (Washington, D.C.: Brookings, 2013).
18. I. Kirsch, H. Braun, and K. Yamamoto, *America's Perfect Storm: Three Forces Changing Our Nation's Future* (Princeton, NJ: Educational Testing Service, 2007).
19. V. Wadhwa, *Why America Is Losing the Global Race to Capture Entrepreneurial Talent* (Philadelphia: Wharton, 2012).
20. Philippe Grandjean, Chemical Brain Drain, "Brain Development Is Uniquely Sensitive," http://braindrain.dk/brain-development-is-uniquely-sensitive. See also, Barbara Demeneix, *Losing Our Minds: How Environmental Pollution Impairs Human Intelligence and Mental Health* (Oxford Series in Behavioral Neuroendocrinology) (New York: Oxford University Press, 2014).
21. P. J. Landrigan et al., "Children's Health and the Environment: A New Agenda for Prevention Research," *Environmental Health Perspectives* 106, Supplement 3 (June 1998): 787–94, https://www.ncbi.nlm.nih.gov/pmc/articles/PMC1533065/.
22. Philippe Grandjean, Chemical Brain Drain, "Brain Development Is Uniquely Sensitive."
23. Landrigan et al., "Children's Health and the Environment."
24. C. Klaassen, ed., *Casarett and Doull's Toxicology: The Basic Science of Poisons,* 5th ed. (New York: McGraw-Hill, 1996).
25. Landrigan et al., "Children's Health and the Environment."
26. Landrigan et al., "Children's Health and the Environment."
27. The Chemical Industry Archive, "Fact #3: The Fetus, Infants, and Children Are Especially Vulnerable to Toxic Substances," http://www.chemicalindustryarchives.org/factfiction/facts/3.asp.
28. United States Food and Drug Administration, "Certain estrogens for oral or parenteral use. Drugs for human use; drug efficacy study implementation," *Fed Regist* 36, no. 217 (1971): 21,537–38, 36 FR 21537, https://www.fda.gov/ohrms/dockets/98fr/cd98191.pdf.
29. Robert Meyers, *D.E.S.: The Bitter Pill* (New York: Seaview/Putnam, 1983).
30. Office of Research on Women's Health, National Institutes of Health, DHHS, "Status of Research on Uterine Fibroids (leiomyomata uteri) at the National Institutes of Health," March 2006, https://dokumen.tips/documents/an-overview-of-uterine-fibroids.html; United States Department of Health and Human Services, Centers for Disease Control and Prevention, "DES Update: For Consumers," https://www.cdc.gov/DES/CONSUMERS/index.html; J. R. Palmer et al., "Urogenital Abnormalities in Men Exposed to Diethylstilbestrol in Utero: A Cohort Study," *Environmental Health* 8, no. 37 (2009): 1–6, https://www.ncbi.nlm.nih.gov/pubmed/19689815.
31. Philippe Grandjean, *Only One Chance: How Environmental Pollution Impairs Brain Development—And How to Protect the Brains of the Next Generation* (New York: Oxford University Press, 2015), 2.

32. That decline was primarily driven by what the authors call a "culling of the least healthy fetuses," resulting in a "horrifyingly large" increase in fetal deaths and miscarriages. Christopher Ingraham, "Flint's Lead-Poisoned Water Had a 'Horrifyingly Large' Effect on Fetal Deaths, Study Finds," *Washington Post,* September 21, 2017, https://www.washingtonpost.com/news/wonk/wp/2017/09/21/flints-lead-poisoned-water-had-a-horrifyingly-large-effect-on-fetal-deaths-study-finds/?utm_term=.217754fa3999.
33. Ibid.
34. Marc Edwards, "Flint Water Crisis Caused by Interrupted Corrosion Control: Investigating 'Ground Zero' Home," *Environmental Science and Technology* 48, no. 1 (2014): 739–46, http://pubs.acs.org/doi/abs/10.1021/acs.est.6b04034.
35. Ibid.
36. Justin Worland, "Air Pollution: This Toxic Pollutant Could Harm the Brain," http://time.com/4480016/air-pollution-health-effects/; Justin Worland, "Preterm Births Linked to Air Pollution Cost Billions in the U.S.," http://time.com/4274355/air-pollution-preterm-birth/.
37. Harriet A. Washington, "Alarming New Diseases among the Urban Poor," *Emerge* 7, no. 6 (April 1996): 22.
38. G. E. Gurri Glass et al., "Association of Chronic Renal Disease, Hypertension, and Infection with a Rat-Borne Hantavirus," *Archives of Virology* 1, Supplement 1 (1990): 69–80.
39. Marc B. Landeet al., "Neurocognitive Function in Children with Primary Hypertension," *The Journal of Pediatrics,* September 2016, https://www.ncbi.nlm.nih.gov/pubmed/27692987.
40. American Academy of Neurology (AAN), "Mom's High Blood Pressure in Pregnancy Could Affect Child's IQ into Old Age," *Science Daily,* October 3, 2012, www.sciencedaily.com/releases/2012/10/121003163631.htm.
41. AAN, "Mom's High Blood Pressure in Pregnancy Could Affect Child's IQ into Old Age."
42. "High Blood Pressure Related Decline in Cognitive Function Affects Adults Young and Old," *ScienceDaily,* October 6, 2004, www.sciencedaily.com/releases/2004/10/041005073315.htm.
43. United States Environmental Protection Agency, "Toxicological Review of Trichloroethylene," September 2011, EPA 635 (R–09/011F). Although TCE's use in dry cleaning was curtailed in the 1950s, it was used as a spot cleaner until 2000 and is still used to clean metal parts.
44. Branch Basics, "HEPA Vacuum: Indoor Air Pollution's Worst Enemy (and How to Choose One for Your Home)," https://branchbasics.com/blog/2015/06/hepa-vacuum-indoor-air-pollutions-worst-enemy/.
45. "'But don't go throwing your trusty vacuum cleaner out so quickly,' says Viviana Temino, MD. She is an assistant professor of allergy and immunology at the University of Miami School of Medicine. 'For a vacuum to do more harm than good, it has to be a really old vacuum cleaner that has

never been cleaned,' she says. 'In general, most vacuums do take up more dust, dirt, and allergens than they release.' HEPA filters are still the way to go, she says: 'They remove more particles than they release back.'" Denise Mann, "Are Vacuum Cleaners Bad for Your Health?," WebMD, January 6, 2012, https://www.webmd.com/allergies/rm-quiz-indoor-allergies.

46. Justin Worland, "Preterm Births Linked to Air Pollution Cost Billions in the U.S.," *Time,* http://time.com/4274355/air-pollution-preterm-birth/.
47. Tom Pelton, "High Sewage Bacteria Levels Found in Back River," *Baltimore Sun,* June 15, 2004, http://articles.baltimoresun.com/2004-06-15/news/0406150094_1_bacteria-public-health-chesapeake-bay.
48. Ibid.; Centers for Disease Control, "About Pfiesteria," July 2004, https://www.cdc.gov/hab/pfiesteria/pdfs/about.pdf.
49. "But the waste contributes to the growing 'dead zone' in the bay, where oxygen is insufficient for life," said Dr. Robert Lawrence, director of the public health school's Center for a Liveable Future. Pelton, "High Sewage Bacteria Levels Found in Back River."
50. David Baker interview with the author, November 27, 2017.
51. Department of Energy Office of Science Biological and Environmental Research, "Methylmercury Sleuths Armed with New Spotlight," https://science.energy.gov/ber/highlights/2017/ber-2017-01-d/.
52. Judith Groch, "Eating Fish During Pregnancy Provides 'Brain Food' for Child," *Medpage Today Ob/Gyn,* February 16, 2007, https://www.medpagetoday.com/obgyn/pregnancy/5075; Joseph Hibbeln, M.D., of the National Institute on Alcohol Abuse and Alcoholism, and colleagues, reported in the February 17 issue of *The Lancet.* In an accompanying comment, Gary Myers, and Philip Davidson, of the University of Rochester wrote that although methylmercury can be neurotoxic, the amount of exposure that constitutes a toxic dose is unknown.
53. Author's telephone interview with Robert Bullard, October 27, 2016.
54. Kathryn R. Mahaffey, Robert P. Clickner, and Rebecca A. Jeffries, "Adult Women's Blood Mercury Concentrations Vary Regionally in the United States: Association with Patterns of Fish Consumption (NHANES 1999–2004)," *Environmental Health Perspectives* 117, no. 1 (2009); see Table 11, http://www.ehponline.org/members/2008/11674/suppl.pdf.
55. Neltner, "When It Comes to Lead, Formula-Fed Infants Get Most from Water and Toddlers from Food, but for Highest Exposed Children the Main Source of Lead Is Soil and Dust."
56. Kathlene Butler et al., "EPA Needs to Provide Leadership and Better Guidance to Improve Fish Advisory Risk Communications," EPA Office of Inspector General Report no. 17-P-0174, April 12, 2017 https://www.epa.gov/sites/production/files/2017-04/documents/_epaoig_20170412-17-p-0174.pdf.
57. Line E. Kirk et al., "Public Health Benefits of Hair-Mercury Analysis and Dietary Advice in Lowering Methylmercury Exposure in Pregnant Women," *Scandinavian Journal of Public Health* 45, no. 4 (April 6, 2017): 444–51, https://www.ncbi.nlm.nih.gov/pubmed/28381203.

58. Vanessa Schipani, "Is Bacon Better for You Than Tilapia?," https://www.factcheck.org/2017/07/bacon-better-tilapia/.
59. Maureen McDonald, "9,000 Flint Residents Sue over Tainted Water," *The Daily Beast,* November 29, 2017, https://www.thedailybeast.com/9000-flint-residents-sue-over-tainted-water.
60. "Data from NHANES III show that, in the southern states, 53%–76% of non-Hispanic blacks (depending on sex and age) compared with 8%–33% of non-Hispanic whites had 25(OH)D concentrations below 50 nmol/L in the winter. The lower 25(OH)D of blacks and other groups with dark skin results primarily from the fact that pigmentation reduces vitamin D production in the skin." Susan S. Harris, "Vitamin D and African Americans," *The Journal of Nutrition* 136, no. 4 (April 2006): 1,126–11, https://www.ncbi.nlm.nih.gov/pubmed/16549493.
61. Jane Kay, "Should Doctors Warn Pregnant Women about Environmental Risks?," *Scientific American,* December 10, 2012, https://www.scientificamerican.com/article/should-doctors-warn-pregnant-women-about-environmental-risks/.
62. Ibid.
63. Ibid.
64. Susan Perry, "Most Ob-Gyns Do Not Warn Pregnant Moms about Environmental Toxins, Survey Finds," *Minnesota Post,* December 13, 2012, https://www.minnpost.com/second-opinion/2012/12/most-ob-gyns-do-not-warn-pregnant-moms-about-environmental-toxins-survey-find.
65. U.S. EPA/Lead Paint Program, "Protect Your Family from Exposures to Lead," https://www.epa.gov/lead/protect-your-family-exposures-lead.
66. Frederica Perera et al., "Prenatal Bisphenol A Exposure and Child Behavior in an Inner-City Cohort," *Environmental Health Perspectives* 120 (April 27, 2012): 1,190–94, https://www.ncbi.nlm.nih.gov/pmc/articles/PMC3440080/.
67. Kay, "Should Doctors Warn Pregnant Women about Environmental Risks?"
68. Jeff Mays, "Harlem Residents Outraged by 'Ghetto' Liquor Store," *New York Times,* November 28, 2011.
69. Tim Scott, "40oz Beats: A Brief History of Malt Liquor in Hip Hop," *Noisey,* November 17, 2015, https://noisey.vice.com/en_au/article/rjxak4/40oz-beats-a-brief-history-of-malt-liquor-in-hip-hop.
70. D. Hackbarth, B. Silvestri, and W. Cosper, "Tobacco and Alcohol Billboards in 50 Chicago Neighborhoods: Market Segmentation to Sell Dangerous Products to the Poor," *Journal of Public Health Policy* 16, no. 2 (1995): 213–30, https://link.springer.com/article/10.2307/3342593?no-access=true#citeas.
71. "Neurobehavioral Disorder Associated with Prenatal Alcohol Exposure (ND-PAE)," *The Diagnostic and Statistical Manual of Mental Disorders (DSM-5)* (New York: American Psychiatric Association, 2013), https://www.mofas.org/resource/the-diagnostic-and-statistical-manual-of-mental-disorders/.
72. Carl C. Bell, "High Rates of Neurobehavioral Disorder Associated with Prenatal Exposure to Alcohol among African Americans Driven by the

Plethora of Liquor Stores in the Community," *Journal of Family Medicine and Disease Prevention* 2, no. 2 (2016): 1.

73. Center for Disease Control and Prevention, "Data & Statistics Prevalence of FASDs," April 16, 2015, https://www.cdc.gov/ncbddd/fasd/facts.html.
74. J. F. Williams and V. C. Smith, "Fetal Alcohol Spectrum Disorders," *Pediatrics* 136, no. 5 (2015): e1,395–1,406.
75. Carl C. Bell, M.D., and Radhika Chimata, M.A., "Prevalence of Neurodevelopmental Disorders among Low-Income African Americans at a Clinic on Chicago's South Side," *Psychiatric Services* 66, no. 5(2015): 539–42, https://ps.psychiatryonline.org/doi/abs/10.1176/appi.ps.201400162?journalCode=ps.
76. West Virginia University School of Public Health, "Alcohol Awareness," http://publichealth.hsc.wvu.edu/alcohol/Effects-on-Society/Gender-and-Ethnic-Differences; http://publichealth.hsc.wvu.edu/.
77. Vivian M. Gonzalez and Monica C. Skewes, "Association of the Firewater Myth with Drinking Behavior among American Indian and Alaska Native College Students," *Psychological Addiction Behavior* 30, no. 8 (2016): 838–49, https://www.ncbi.nlm.nih.gov/pubmed/27736147.
78. By an epigenetic mechanism involving a nicotinic acetylcholine receptor. Carl C. Bell and Jessie Aujla, "Prenatal Vitamins Deficient in Recommended Choline Intake for Pregnant Women," *American Journal of Family Medicine and Disease Prevention* 173, no. 5 (2016): 509–16.
79. "Morgan Fawcett on Living with FASD," https://www.youtube.com/watch?v=K0VrkLQfkFg.
80. Office of Juvenile Justice and Delinquency, "January 2017 Office of Juvenile Justice and Delinquency Prevention Report," https://www.ncjrs.gov/pdffiles1/ojjdp/fs200102.pdf.
81. Includes "Living in a stable, nurturing home for more than 72 percent of their life; Being diagnosed before the age of 6 years; Never having experienced violence against themselves; Staying in a living situation for an average of more than 2.8 years; Experiencing a good quality of home life; Having applied for and been eligible for developmental disability services; Having a diagnosis of FAS (rather than partial *fetal alcohol* syndrome (PFAS)."
82. American Public Health Association (APHA), "Advertising and Promotion of Alcohol and Tobacco Products to Youth," American Public Health Association Policy Number 9213, January 1, 1992, https://www.apha.org/policies-and-advocacy/public-health-policy-statements/policy-database/2014/07/29/10/58/advertising-and-promotion-of-alcohol-and-tobacco-products-to-youth.
83. Ibid.
84. CDC Highlights, "Marketing Cigarettes to Women," 2001, https://www.cdc.gov/tobacco/data_statistics/sgr/2001/highlights/marketing/index.html.
85. According to the CDC, "A study of 111 women's magazines in 17 European countries in 1996–1997 found that 55% of the magazines that

responded accepted cigarette advertisements, and only 4 had a policy of voluntarily refusing it. Only 31% of the magazines had published an article of one page or more on smoking and health in the previous 12 months. Magazines that accepted tobacco advertisements seem less likely to give coverage to smoking and health issues." Ibid.

86. Harriet A. Washington, "Burning Love: Big Tobacco Takes Aim at LGBT Youths," *American Journal of Public Health* 92, no. 7 (2002): 1,086–95, http://ajph.aphapublications.org/doi/full/10.2105/AJPH.92.7.1086.
87. U.S. Department of Health and Human Services/CDC, "Smoking and Pregnancy," based on the 2014 Surgeon General's Report, https://www.cdc.gov/features/pregnantdontsmoke/pregnantdontsmoke.pdf.
88. Kay, "Should Doctors Warn Pregnant Women about Environmental Risks?"
89. M. Borum, "A Comparison of Smoking Cessation Efforts in African Americans by Resident Physicians in a Traditional and Primary Care Internal Medicine Residency," *Journal of the National Medical Association* 92, no. 3 (March 2000): 131–35.
90. Linda Villarosa, "Why America's Black Mothers and Babies Are in a Life-or-Death Crisis," *New York Times,* April 11, 2018.
91. Zoë Carpenter, "What's Killing America's Black Infants?," *The Nation,* February 15, 2017, https://www.thenation.com/article/whats-killing-americas-black-infants/.
92. Tamar Lewin, "Implanted Birth Control Device Renews Debate over Forced Contraception," *New York Times,* January 10, 1991, https://www.nytimes.com/1991/01/10/us/implanted-birth-control-device-renews-debate-over-forced-contraception.html.
93. Ibid.
94. Harriet A. Washington, *Medical Apartheid: The Dark History of Medical Experimentation on Black Americans from Colonial Times to the Present* (New York: Doubleday, 2007).
95. "A Woman's Rights: Part 4; Slandering the Unborn," https://www.nytimes.com/interactive/2018/12/28/opinion/crack-babies-racism.html.
96. Madeline Ostrander, "What Poverty Does to the Young Brain," *The New Yorker,* June 4, 2015, https://www.newyorker.com/tech/elements/what-poverty-does-to-the-young-brain.
97. Linda C. T. Mayes, "The Problem of Prenatal Cocaine Exposure: A Rush to Judgment," *Journal of the American Medical Association* 267, no. 267 (1992): 406; Barry Zuckerman and Frank Deborah, "'Crack Kids': Not Broken," *Pediatrics* 89 (1992): 337; Robert Mathias, "Developmental Effects of Prenatal Drug Exposure May Be Overcome by Postnatal Environment," *NIDA Notes* (1992): 14.
98. Richard Hindmarsh, Barbara Prainsack, eds., *Genetic Suspects: Global Governance of Forensic DNA Profiling and Databasing* (Cambridge: Cambridge University Press, 2010), 70.
99. Carpenter, "What's Killing America's Black Infants?"
100. Demeneix, *Toxic Cocktail.*

101. Barbara Demeneix, *Losing Our Minds.*
102. Jo Jones and William Mosher, "Fathers' Involvement with Their Children: United States, 2006–2010," Division of Vital Statistics National Health Statistics Report Number 71n, December 20, 2013, https://pdfs.semanticscholar.org/9140/a75bcccc5b4c00a0f1b077a56597130ae983.pdf. See also, Howard Dubowitz et al., "Low-Income African American Fathers' Involvement in Children's Lives: Implications for Practitioners," *Journal of Family Social Work* 10, no. 1 (2006): 25–41, http://www.tandfonline.com/doi/abs/10.1300/J039v10n01_02?journalCode=wfsw20.

Chapter 5: Bugs in the System: How Microbes Sap U.S. Intelligence

1. Peter Hotez, "'Justice for All' Should Embrace Black America's Neglected Health Disparities," *Huffington Post,* December 23, 2014, updated February 21, 2015, https://www.huffingtonpost.com/peter-hotez-md-phd/justice-for-all-should-em_b_6367290.html.
2. Christopher Eppig, "Why Is Average IQ Higher in Some Places? A Surprising Theory about Global Variations in Intelligence," *Scientific American,* September 6, 2011, https://www.scientificamerican.com/article/why-is-average-iq-higher-in-some-places/.
3. Liz Szabo, "WHO: Sexual Transmission of Zika More Common Than Thought," *Chicago Tribune,* March 8, 2016, http://www.chicagotribune.com/news/nationworld/ct-zika-sexually-transmitted-20160308-story.html.
4. Donald G. McNeil Jr., "Zika May Increase Risk of Mental Illness, Researchers Say," *New York Times,* February 18, 2016, https://www.nytimes.com/2016/02/23/health/zika-may-increase-risk-of-mental-illness-researchers-say.html.
5. Ibid.
6. Harriet Washington, "Zika May Bring a Wave of Mental Health Problems in Future Years," *New Scientist,* April 19, 2016, https://www.newscientist.com/article/2084870-zika-may-bring-a-wave-of-mental-health-problems-in-future-years/.
7. WHO Media Centre, "Zika Virus Fact Sheet," updated February 2016, http://www.who.int/mediacentre/factsheets/zika/en/.
8. Jernej Mlakar, M.D., et al., "Zika Virus Associated with Microcephaly," *New England Journal of Medicine* 374 (2016): 951–58, doi:10.1056/NEJMoa1600651, https://www.nejm.org/doi/pdf/10.1056/NEJMoa1600651.
9. Andy Coghlan, "Whole Zika Genome Recovered from Brain of Baby," *New Scientist,* February 10, 2016, https://www.newscientist.com/article/2077091-whole-zika-genome-recovered-from-brain-of-baby-with-microcephaly/.
10. "Diseases and Conditions: Microencephaly," Mayo Clinic, http://www.mayoclinic.org/diseases-conditions/microcephaly/basics/definition/CON-20034823.

11. "Microcephaly Symptoms and Causes," Boston's Children's Hospital, http://www.childrenshospital.org/conditions-and-treatments/conditions/microcephaly/symptoms-and-causes.
12. Washington, "Zika may bring a wave of mental health problems in future years."
13. McNeil, "Zika May Increase Risk of Mental Illness, Researchers Say."
14. Ibid.
15. Centers for Disease Control and Prevention, "About Rubella," http://www.cdc.gov/rubella/about/index.html.
16. M. A. Holliday, "Body Composition and Energy Needs During Growth," in *Human Growth: A Comprehensive Treatise,* eds. F. Falkner and J. M. Tanner (New York: Plenum, 1986): 2,101–17.
17. Harriet A. Washington, "The Well Curve: Tropical Diseases Are Undermining Intellectual Development in Countries with Poor Health Care—And They're Coming Here Next," *American Scholar,* September 7, 2015, https://theamericanscholar.org/the-well-curve/#.XEDdKzbGLzI.
18. McNeil, "Zika May Increase Risk of Mental Illness, Researchers Say."
19. W. Kleber de Oliveira et al., "Increase in Reported Prevalence of Microcephaly in Infants Born to Women Living in Areas with Confirmed Zika Virus Transmission During the First Trimester of Pregnancy—Brazil, 2015," *Morbidity and Mortality Weekly Report* 65 (2016): 242–47. "Originally, doctors in Brazil believed that infections in the first trimester were the most dangerous, because mothers who gave birth to babies with microcephaly were usually infected then." Donald G. McNeil Jr., Catherine Saint Louis, and Nicholas St. Fleur, "Short Answers to Hard Questions about Zika Virus," *New York Times,* March 18, 2016, https://www.nytimes.com/interactive/2016/health/what-is-zika-virus.html.
20. Donald G. McNeil Jr. and Catherine Saint Louis, "Two Studies Strengthen Links between the Zika Virus and Serious Birth Defects," *New York Times,* March 4, 2016, https://www.nytimes.com/2016/03/05/health/zika-virus-microcephaly-fetus-birth-defects.html.
21. Guillain-Barré Syndrome Fact Sheet, National Institute of Neurological Disorders and Stroke, http://www.ninds.nih.gov/disorders/gbs/detail_gbs.htm.
22. "Stupidity or Hookworm?," *Virginia Health Bulletin* 26 (1934): 13–16.
23. Ed Pilkington, "Hookworm, a Disease of Extreme Poverty, Is Thriving in the U.S. South. Why?," *The Guardian,* September 5, 2017.
24. Oscar Rickett, "Tropical Diseases Are Keeping Americans in Poverty," *Vice,* September 29, 2014, https://www.vice.com/en_us/article/wd47vx/neglected-tropical-diseases-are-keeping-americans-in-poverty-930.
25. Peter Hotez et al., "Texas and Mexico: Sharing a Legacy of Poverty and Neglected Tropical Diseases," *PLOS* 6, no. 3 (2012): e1,497.
26. Washington, "The Well Curve."
27. Ibid.

28. Wilma A. Stolk et al., "Between-Country Inequalities in the Neglected Tropical Disease Burden in 1990 and 2010, with Projections for 2020," *PLOS Neglected Tropical Diseases* 8, no. 9 (2014), https://www.ncbi.nlm.nih.gov/pmc/articles/PMC4865216/.
29. "To Breed, or Not to Breed: A Fearsome Outbreak Has Triggered a Debate about Birth Control," *The Economist,* January 30, 2016, https://www.economist.com/news/americas/21689618-fearsome-outbreak-has-triggered-debate-about-birth-control-breed-or-not-breed.
30. A. S. Fauci and D. M. Morens, "Zika Virus in the Americas—Yet Another Arbovirus Threat," *New England Journal of Medicine* 374 (January 13, 2016): 601–4, http://www.nejm.org/doi/full/10.1056/NEJMp1600297.
31. Author's telephone interview with Christopher Eppig, June 15, 2016.
32. Mireia Valles-Colomer et al., "The neuroactive potential of the human gut microbiota in quality of life and depression," *Nature Microbiology,* February 4, 2019, https://www.nature.com/articles/s41564-018-0337-x.
33. Harriet A. Washington, *Infectious Madness: The Surprising Science of How We "Catch" Mental Illness* (New York: Little, Brown, 2015).
34. Christopher Eppig, Corey L. Fincher, and Randy Thornhill, "Parasite prevalence and the worldwide distribution of cognitive ability," *Proceedings of the Biolological Sciences* 277; 1701 (December 22, 2010): 3,801–8, https://www.ncbi.nlm.nih.gov/pmc/articles/PMC2992705/.
35. J. M. Wicherts et al., "Raven's Test Performance of Sub-Saharan Africans: Mean Level, Psychometric Properties and the Flynn Effect," *Learning and Individual Differences* (2010), http://citeseerx.ist.psu.edu/viewdoc/download;jsessionid=A2F9B41C0A9139144AC0D9676C14DC50?doi=10.1.1.729.9100&rep=rep1&type=pdf; Richard Lynn and Tatu Vanhanen, *IQ and the Wealth of Nations* (Westport, CT: Praeger, 2002).
36. Christopher Eppig, Corey L. Fincher, and Randy Thornhill, "Parasite Prevalence and the Worldwide Distribution of Cognitive Ability," *Proceedings of the Royal Society B,* June 30, 2010, http://rspb.royalsocietypublishing.org/content/277/1701/3801.
37. Eppig, "Why Is Average IQ Higher in Some Places?"
38. Ian Sample, "Lower IQs found in disease-rife countries, scientists claim," *The Guardian,* June 29, 2010, https://www.theguardian.com/science/2010/jun/30/disease-rife-countries-low-iqs.
39. Eppig, "Why Is Average IQ Higher in Some Places?"
40. Ibid.
41. C. Hassall and T. N. Sherratt, "Statistical inference and spatial patterns in correlates of IQ," *Intelligence* 39, no. 5 (2011): 303–10, https://www.sciencedirect.com/science/article/pii/S0160289611000572.
42. Debora MacKenzie, "Link Found between Infectious Disease and IQ," *New Scientist,* June 30, 2010, https://www.newscientist.com/article/mg20727670-301-link-found-between-infectious-disease-and-iq/.
43. John J. Goldman, "Men's Death Rate Higher in Harlem Than Bangladesh," *Los Angeles Times,* January 18, 1990, http://articles.latimes.com/1990-01-18/news/mn-336_1_death-rate.

44. "Examining the U.S. Public Health Response to the Zika Virus," Testimony of Peter Hotez, M.D., Ph.D., Dean, National School of Tropical Medicine, Baylor College of Medicine before the Subcommittee on Oversight and Investigations Committee on Energy and Commerce United States House of Representatives, March 2, 2016, docs.house.gov/.../HHRG-114-IF02-Wstate-HotezP-20160302.pdf.
45. Suad Kapetanovic et al., "Relationships between Markers of Vascular Dysfunction and Neurodevelopmental Outcomes in Perinatally HIV-Infected Youth," *AIDS* 24, no. 10 (2010): 1,481–91, https://www.ncbi.nlm.nih.gov/pubmed/20539091.
46. Ibid.
47. Peter Hotez, "'Justice for All': Should Embrace Black America's Neglected Health Disparities," *HuffPost,* https://www.huffingtonpost.com/peter-hotez-md-phd/justice-for-all-should-em_b_6367290.html.
48. "Examining the U.S. Public Health Response to the Zika Virus."
49. U.S. Environmental Protection Agency, "DDT—A Brief History and Status," https://www.epa.gov/ingredients-used-pesticide-products/ddt-brief-history-and-status; Debora MacKenzie, "America's Hidden Epidemic of Tropical Diseases," *New Scientist,* December 11, 2013, https://www.newscientist.com/article/mg22029473-200-americas-hidden-epidemic-of-tropical-diseases/.
50. Peter Hotez, "America's 'New' Diseases of Poverty," *HuffPost,* March 10, 2014, https://www.huffingtonpost.com/peter-hotez-md-phd/americas-new-diseases-of-poverty_b_4557630.html.
51. Rumona Dicksonet al., "Effects of Treatment for Intestinal Helminth Infection on Growth and Cognitive Performance in Children: Systematic Review of Randomised Trials," *British Medical Journal* 320, no. 7251 (2000): 1,697–1,701, doi:10.1136/bmj.320.7251.1697.
52. Cited in Olga Khazan, "Why Are So Many Middle-Aged White Americans Dying?," *The Atlantic,* January 29, 2016, https://www.theatlantic.com/health/archive/2016/01/middle-aged-white-americans-left-behind-and-dying-early/433863/.
53. Washington, *Infectious Madness.*
54. Washington, "The Well Curve."
55. Brazil's average IQ is 87, a ranking it shares with nine other countries, all in the developing world. Lynn and Vanhanen, *IQ and the Wealth of Nations,* 200; "Chagas disease (American trypanosomiasis)," WHO Fact sheet, no. 340, updated March 2015, http://www.who.int/mediacentre/factsheets/fs340/en/.
56. MacKenzie, "America's Hidden Epidemic of Tropical Diseases."
57. Ibid.
58. Fully 80 percent of the world's 50 million people who are affected by epilepsy live in low- and lower-middle-income countries where tapeworms abound, and the United States is now one of these countries. WHO, "Taeniasis/cysticercosis Fact sheet," updated April 2016, http://www.who.int/mediacentre/factsheets/fs376/en/.

59. Hotez et al., "Texas and Mexico."
60. Washington, "The Well Curve," 12–23.
61. Centers for Disease Control and Prevention, "Fact Sheet: Toxocariasis," Centers for Disease Control and Prevention, Division of Parasitic Diseases, updated January 19, 2009, http://www.cdc.gov/ncidod/dpd/parasites/toxocara/factsht_toxocara.htm.
62. Peter Hotez, "Neglected Parasitic Infections and Poverty in the United States," *PLOS Neglected Tropical Diseases* 8, no. 9 (2014): 2, http://journals.plos.org/plosntds/article?id=10.1371/journal.pntd.0003012; Hotez, "'Justice for All.'"
63. Dana M. Woodhall, Mark L. Eberhard, and Monica E. Parise, "Neglected Parasitic Infections in the United States: Toxocariasis," *American Journal of Tropical Medicine and Hygiene* 9, no. 5 (May 7, 2014): 810–13, https://www.ncbi.nlm.nih.gov/pmc/articles/PMC4015569/.
64. Hotez, "'Justice for All.'"
65. MacKenzie, "America's Hidden Epidemic of Tropical Diseases."
66. P. L. Cummings et al., "Trends, Productivity Losses, and Associated Medical Conditions among Toxoplasmosis Deaths in the United States, 2000–2010," *American Journal of Tropical Medical Hygiene* 91, no. 5 (November 2014): 959–64, doi:10.4269/ajtmh.14-0287.
67. Paul R. Torgerson and Pierpaolo Mastroiacovo, "The Global Burden of Congenital Toxoplasmosis: A Systematic Review," *Bulletin of the World Health Organization* 91 (2013):01–8, doi: http://dx.doi.org/10.2471/BLT.12.111732; G. Schmid, "Trichomoniasis Treatment in Women: RHL Commentary," revised July 28, 2003, The WHO Reproductive Health Library, http://apps.who.int/rhl/rti_sti/gscom/en/.
68. MacKenzie, "America's Hidden Epidemic of Tropical Diseases."
69. National Center for Immunizations and Respiratory Diseases, Division of Viral Diseases, "Helping Children with Congenital CMV," December 8, 2017, https://www.cdc.gov/features/cytomegalovirus/index.html.
70. Mayo Foundation for Medical Education and Research, "Cytomegalovirus (CMV) infection," CON-2024573, http://www.mayoclinic.org/diseases-conditions/cmv/home/ovc-20315443.
71. Ibid.
72. "Examining the U.S. Public Health Response to the Zika Virus."
73. Justin Gillis, "In Zika Epidemic, a Warning on Climate Change," *New York Times,* February 20, 2016.
74. Rick Noack, "Europe to America: Your Love of Air Conditioning Is Stupid," *Washington Post,* July 26, 2015, https://www.washingtonpost.com/news/worldviews/wp/2015/07/22/europe-to-america-your-love-of-air-conditioning-is-stupid/?utm_term=.e3e8e070f62d.
75. "[In] the 1960s several South American countries—including Brazil—eliminated [*Aedes aegypti*] by spraying with DDT and urging households to get rid of breeding sites. Unfortunately the mosquitoes survived in a few locations, and after the development of a yellow fever vaccine, campaigns dwindled." Clare Wilson, "7 Ways the War on Zika Mosquitoes Could Be

Won," *New Scientist,* February 2, 2016, https://www.newscientist.com/article/2076078-7-ways-the-war-on-zika-mosquitoes-could-be-won/.

76. "Examining the U.S. Public Health Response to the Zika Virus."
77. Hotez, "Neglected Parasitic Infections and Poverty in the United States."
78. Benjamin K. Sovacool, "Don't Let Disaster Recovery Perpetuate Injustice," *Nature* 549 (2017): 433, https://www.nature.com/news/don-t-let-disaster-recovery-perpetuate-injustice-1.22668.
79. Ibid.
80. Washington, "The Well Curve."
81. "Tracking Dust across the Atlantic," Earth Observatory NASA map, https://earthobservatory.nasa.gov/IOTD/view.php?id=81864.
82. Robert H. Yolken et al., "Chlorovirus ATCV-1 Is Part of the Human Oropharyngeal Virome and Is Associated with Changes in Cognitive Functions in Humans and Mice," *PNAS* 111, no. 45 (2014): 16,106–11, doi: 10.1073/pnas.1418895111; Elisabeth Pennisi, "Algal Virus Found in Humans, Slows Brain Activity," *Science,* October 27, 2014, http://news.sciencemag.org/biology/2014/10/algal-virus-found-humans-slows-brain-activity.
83. George Dvorsky, "Scientists Discover a Virus That Makes Humans Less Intelligent," November 10, 2014, *io9,* https://io9.gizmodo.com/a-virus-that-makes-us-less-intelligent-has-been-discove-1656793377; Kevin Loriaa, "Virus Found in Lakes May Be Literally Changing the Way People Think," *Business Insider,* October 30, 2014, http://www.businessinsider.com/algae-virus-may-be-changing-cognitive-ability-2014-10.
84. Robert H. Yolken et al., "Reply to Kjartansdóttir et al.: Chlorovirus ATCV-1 Findings Not Explained by Contamination," *Proceedings of the National Academy of Sciences* 112, no. 9 (March 2015): E927, https://www.ncbi.nlm.nih.gov/pmc/articles/PMC4352770/.
85. Meghan Cartwright "Don't Forget Eating These Clams Can Cause Amnesia," *Slate,* June 30, 2015, http://www.slate.com/blogs/wild_things/2015/06/30/razor_clams_poisoning_can_cause_diarrhea_amnesia_and_death.html.
86. Ibid.
87. Frances M. Van Dolah, "Marine Algal Toxins: Origins, Health Effects, and Their Increased Occurrence," *Environmental Health Perspectives* 108, Supplement (2000): 133–41, https://www.ncbi.nlm.nih.gov/pmc/articles/PMC1637787/.
88. National Institutes of Health, "Creutzfeldt-Jakob Disease Fact Sheet," May 10, 2017, http://www.ninds.nih.gov/disorders/cjd/detail_cjd.htm.
89. Centers for Disease Control, "Powassan Virus Symptoms and Treatment," February 9, 2015, https://www.cdc.gov/powassan/symptoms.html; Ashley May, "Protect Yourself from Infected Ticks Carrying Life-Threatening Powassan Virus," *USA Today,* May 3, 2017, https://www.usatoday.com/story/news/nation-now/2017/05/03/powassan-virus-how-protect-yourself-infected-ticks/310009001/.

90. "How Safe Is Deet?," *Consumer Reports,* May 8, 2018, https://www.consumerreports.org/insect-repellent/how-safe-is-deet-insect-repellent-safety/.
91. "Study Linking Beneficial Bacteria to Mental Health Makes Top 10 List for Brain Research," *CU Boulder Today,* January 5, 2017, https://www.colorado.edu/today/2017/01/05/study-linking-beneficial-bacteria-mental-health-makes-top-10-list-brain-research.
92. Christie Nicholson, "Soil Bacteria Might Increase Learning," *Scientific American,* May 24, 2010, https://www.scientificamerican.com/podcast/episode/soil-bacteria-might-increase-learni-10-05-24/.
93. S. Z. Szatmari and P. J. Whitehouse, "Vinpocetine for Cognitive Impairment and Dementia," *Cochrane Database System Review* 1 (2003): CD003119, https://www.ncbi.nlm.nih.gov/pubmed/12535455.
94. A. O. Ogunrin, "Effect of Vinpocetine (Cognitol) on Cognitive Performances of a Nigerian Population," *Annals of Medical Health Science Research* 4, no. 4 (July–August 2014): 654–61, https://www.ncbi.nlm.nih.gov/pubmed/25221724.
95. Other names by which it is known include AY-27255, Eburnamenine-14-carboxylic acid, Ethyl Apovincaminate, Ethylapovincaminoate, Ethyl Ester, RGH-4405, TCV-3b, Vinpocetin, Vinpocetina, and Vinpocétine.

Chapter 6: Taking the Cure: What Can You Do, Now?

1. M. B. Pell and Joshua Schneyer, "The Thousands of U.S. Locales Where Lead Poisoning Is Worse Than in Flint," https://www.reuters.com/investigates/special-report/usa-lead-testing/.
2. Robert D. Bullard, "Environmental Justice for All," *Issues & Views,* http://www.uky.edu/~tmute2/GEI-Web/password-protect/GEI-readings/Bullard-Environmental%20justice%20for%20all.pdf.
3. Susan Edelman, "DOE Officials Evasive about Lead-Tainted Fountains in Schools," *New York Post,* September 8, 2018, https://nypost.com/2018/09/08/doe-officials-evasive-about-lead-tainted-fountains-in-schools; Greg B. Smith, "Top Managers Resign, Another Demoted after NYCHA Lied about Lead Paint Inspections," *New York Daily News,* November 17, 2017, https://www.nydailynews.com/new-york/managers-resign-demoted-nycha-lead-paint-scandal-article-1.3640566; Selim Agar, "DOE Claims Schools Have Never Had a Case of Lead Poisoning," *New York Post,* April 18, 2007, https://nypost.com/2017/04/18/doe-claims-schools-have-never-had-a-case-of-lead-poisoning/.
4. Kirsten McCulloch, *Less Toxic Living: How to Reduce Your Everyday Exposure to Toxic Chemicals—An Introduction for Families* (Canberra, Australia: Green Gables Press, 2013).
5. Ibid., 131–32.
6. "How Safe Is Your Drinking Water?," *Consumer Reports,* March 21, 2016, https://www.consumerreports.org/water-filters/how-safe-is-your-drinking-water/.

7. Donald L. Chi, personal communication, and Donald L. Chi et al., "Caregivers' understanding of fluoride varnish: implications for future clinical strategies and research on preventive care decision making," *Journal of Public Health Dentistry* (August 2018).
8. M. Bashash et al., "Prenatal Fluoride Exposure and Cognitive Outcomes in Children at 4 and 6–12 Years of Age in Mexico," *Environmental Health Perspectives* 125, no. 9 (2017): 097017-1-12, https://ehp.niehs.nih.gov/ehp655/.
9. Michelle Manchir, "Lawsuit Seeking Fluoridation Ban Moves Forward," *ADA News,* January 5, 2018, https://www.ada.org/en/publications/ada-news/2018-archive/january/lawsuit-seeking-fluoridation-ban-moves-forward?source=promospots&medium=ADAFluorideRotator&content=Lawsuit.
10. Ashley May, "Children are using an unhealthy amount of toothpaste, CDC warns," *USA Today,* February 4, 2019, https://www.usatoday.com/story/news/health/2019/02/04/children-using-too-much-toothpaste-unhealthy-cdc/2766121002/.
11. W. J. Rogan and B. C. Gladen, "Breast-feeding and Cognitive Development," *Early Human Development* 31, no. 3 (1993): 181–93.
12. G. Mehta and N. Sheron, "No Safe Level of Alcohol Consumption—Implications for Global Health." *Journal of Hepatology* (2019), https://www.journal-of-hepatology.eu/article/S0168-8278(18)32640-0/pdf.
13. Ashley May, "These Baby Foods and Formulas Tested Positive for Arsenic, Lead and BPA in New Study," *USA Today,* October 25, 2017.
14. Kathleen Zelman, "The Vital Role of Food Preservatives," *Food and Nutrition,* February 27, 2017, https://foodandnutrition.org/march-april-2017/vital-role-food-preservatives.
15. USDA, "Complete Guide to Home Canning: Principles of Home Canning," http://nchfp.uga.edu/publications/usda/GUIDE01_HomeCan_rev0715.pdf.
16. Andy Coghlan, "Extreme Mercury Levels Revealed in Whalemeat," *New Scientist,* June 6, 2002, https://www.newscientist.com/article/dn2362-extreme-mercury-levels-revealed-in-whalemeat//.
17. D. H. Phua, A. Zosel, and K. Heard, "Dietary Supplements and Herbal Medicine Toxicities—When to Anticipate Them and How to Manage Them," *International Journal of Emergency Medicine* 2, no. 2 (June 2009): 69–76.
18. David M. Eisenberg et al., "Unconventional Medicine in the United States—Prevalence, Costs, and Patterns of Use," *New England Journal of Medicine* 328 (1993): 246–52.
19. "Aconite Overview," WebMD, https://www.webmd.com/vitamins/ai/ingredientmono-609/aconite.
20. "Pure and Highly Concentrated Caffeine," FDA, April 13, 2018, https://www.fda.gov/Food/DietarySupplements/ProductsIngredients/ucm460095.htm.

21. "Toxic Hepatitis," Mayo Clinic, https://www.mayoclinic.org/diseases-conditions/toxic-hepatitis/symptoms-causes/syc-20352202.
22. M. Roulet et al., "Hepatic Veno-occlusive Disease in Newborn Infant of a Woman Drinking Herbal Tea," *Journal of Pediatrics* 112 (1988): 433–36.
23. M. L. Yeong et al., "Hepatic Veno-occlusive Disease Associated with Comfrey Ingestion," *Journal of Gastroenterology and Hepatology* 5, no. 2 (1990): 211–14, doi:10.1111/j.1440-1746.1990.tb01827.x. PMID 2103401.
24. Amitava Dasgupta and Jorge L. Sepulveda, "Effect of Herbal Remedies on Clinical Laboratory Tests," in *Accurate Results in the Clinical Laboratory,* eds. Amitava Dasgupta and Jorge L. Sepulveda (New York: Elsevier, 2013), 75–92, https://www.sciencedirect.com/topics/pharmacology-toxicology-and-pharmaceutical-science/teucrium.
25. Joerg Gruenwald, *PDR for Herbal Medicines* (Montvale, NJ: Thomson, 2000), 1-1, 0-13, 64, 428, 584, 974.
26. "Dangerous Supplements: Still at Large," *Consumer Reports,* May 2004, https://www.consumerreports.org/video/view/healthy-living/drugs-medication/605132728001/dangerous-dietary-supplements/.
27. Amanda MacMillan, "These 15 Supplement Ingredients Carry Serious Health Risks, According to a New Report," *Health,* July 27, 2016, http://www.health.com/nutrition/supplement-ingredients-to-avoid.
28. Andrew Chadwick et al., "Accidental Overdose in the Deep Shade of Night: A Warning on the Assumed Safety of 'Natural Substances,'" *BMJ Case Reports* (2015), doi:10.1136/bcr-2015-209333.
29. Lesley Braun and Marc Cohen, "Introduction to Complementary Medicine," in *Herbs and Natural Supplements, Volume 1: An Evidence-Based Guide* (Sydney, Australia: Churchill Livingstone, 2014), 9, https://books.google.com/books?id=xVqmBgAAQBAJ&pg=PA9&lpg=PA9&dq=an+analysis+of+2,609+herbal+samples&source=bl&ots=OMwt7VhW0R&sig=NULGGbWAXbMHXf3SuVECyNzIfo0&hl=en&sa=X&ved=0ahUKEwiCrLjv1LDbAhWjtlkKHZJmBrQQ6AEIMTAB#v=onepage&q=an%20analysis%20of%202%2C609%20herbal%20samples&f=false.
30. Ghazal Mortazavi and S. M. J. Mortazavi, "Increased Mercury Release from Dental Amalgam Restorations after Exposure to Electromagnetic Fields as a Potential Hazard for Hypersensitive People and Pregnant Women," *Reviews on Environmental Health* 30, no. 4 (2015): 287–92.
31. WHO, "Mercury and Health," January 2016, https://web.archive.org/web/20161120171147/http:/www.who.int/mediacentre/factsheets/fs361/en/.
32. Simon Cotton, "Dimethylmercury and Mercury Poisoning," April 17, 2012 at WebCite—University of Bristol web page documenting poisoning death, retrieved December 9, 2006, http://www.chm.bris.ac.uk/motm/dimethylmercury/dmmh.htm.
33. European Parliament, "Export-Bn of Mercury and Mercury Compounds from the EU by 2011," press release, May 21, 2008, http://www.europarl

.europa.eu/news/expert/infopress_page/064-29478-140-05-21-911-20080520IPR29477-19-05-2008-2008-false/default_en.htm.
34. "Lead Test Kits," *Consumer Reports,* May 2016, https://www.consumerreports.org/cro/lead-test-kits/buying-guide.htm.

Chapter 7: A Wonderful Thing to Save: How Communities Can Unite to Preserve Brainpower

1. Environmental Protection Agency, "EPA Bans PCB Manufacture; Phases Out Uses," EPA Press Release, April 19, 1979, last updated August 8, 2016, https://archive.epa.gov/epa/aboutepa/epa-bans-pcb-manufacture-phases-out-uses.html.
2. "Carolinians Angry over PCB Landfill," *New York Times,* August 11, 1982, http://www.nytimes.com/1982/08/11/us/carolinians-angry-over-pcb-landfill.html.
3. Ibid.
4. Ibid.; Deborah Ferruccio, "Birth of a Movement—Environmental Justice and Pollution Prevention," NCPCB Archives, www.ncpcbarchives.com/?page_id=144.
5. Ibid.
6. "Carolinians Angry over PCB Landfill."
7. "Dumping on the Poor," *Washington Post,* October 12, 1982, https://www.washingtonpost.com/archive/politics/1982/10/12/dumping-on-the-poor/bb5c9b8c-528a-45b0-bd10-874da288cd59/.
8. General Accounting Office, "Siting of Hazardous Waste Landfills and Their Correlation with Racial and Economic Status of Surrounding Communities," GAO//RCED-83-168, June 1, 1983, http://archive.gao.gov/d48t13/121648.pdf.
9. Union of Concerned Scientists, "Pollution Affects Americans Unequally," interview with Robert Bullard, *Catalyst,* Spring 2017, https://www.ucsusa.org/sp17-inquiry-robert-bullard#.WpTOLIJOnYo.
10. Lena Williams, "Race Bias Found in Location of Toxic Dumps," *New York Times,* April 16, 1987, http://www.nytimes.com/1987/04/16/us/race-bias-found-in-location-of-toxic-dumps.html?mcubz=2.
11. Vann R. Newkirk II, "Fighting Environmental Racism in North Carolina," *The New Yorker,* January 16, 2016, https://www.newyorker.com/news/news-desk/fighting-environmental-racism-in-north-carolina.
12. G. Holland, "A Study of Ethnic Markets [by R. J. Reynolds marketing officials]," September 1969, cited in Harriet A. Washington, "Burning Love: Big Tobacco Takes Aim at LGBT Youths, *American Journal of Public Health* 92, no. 7 (2002): 1,086–95, http://ajph.aphapublications.org/doi/full/10.2105/AJPH.92.7.1086.
13. Mireia Gascon, "Mental Health Benefits of Long-Term Exposure to Residential Green and Blue Spaces: A Systematic Review, *International Journal of Environmental Research and Public Health* 12 (2015): 4,354–79.

14. Ibid.
15. Michelle C. Kondo et al., "Urban Green Space and Its Impact on Human Health," *International Journal of Environmental Research and Public Health* 15 (2018): 445, doi:10.3390/ijerph15030445.
16. S. W. Porges, "The Polyvagal Perspective," *Biological Psychiatry* 74 (2007): 116–43.
17. J. F. Brosschot, B. Verkuilos, and J. F. Thayer, "Generalized Unsafety Theory of Stress: Unsafe Environments and Conditions, and the Default Stress Response," *International Journal of Environmental Research and Public Health* 15(3) (2018): E464, doi:10.3390/ijerph15030464; https://www.ncbi.nlm.nih.gov/pubmed/29518937.
18. J. C. Motzkin et al., "Ventromedial Prefrontal Cortex Is Critical for the Regulation of Amygdala Activity in Humans," *Biological Psychiatry* 77 (2015): 276–84.
19. S. F. Maier, "Behavioral Control Blunts Reactions to Contemporaneous and Future Adverse Events: Medial Prefrontal Cortex Plasticity and a Corticostriatal Network," *Neurobiological Stress* 1 (2015): 12–22; S. F. Maier, "Medial Prefrontal Cortex Determines How Stressor Controllability Affects Behavior and Dorsal Raphe Nucleus," *Nature Neuroscience* 8 (2005): 365–71.
20. K. D. Kochanek et al., "How Did Cause of Death Contribute to Racial Differences in Life Expectancy in the United States in 2010?," *NCHS Data Brief* 125 (2013): 1–8.
21. R. Clark et al., "Racism as a Stressor for African Americans—A Biopsychosocial Model," *American Psychologist* 54 (1999): 805–16.
22. C. M. Dolezsar et al., "Perceived Racial Discrimination and Hypertension: A Comprehensive Systematic Review," *Health Psychology* 33 (2014): 20–34.
23. S. Völker and T. Kistemann, "The Impact of Blue Space on Human Health and Well-Being—Salutogenetic Health Effects of Inland Surface Waters: A Review," *International Journal of Hygiene and Environmental Health* 214 (2011): 449–60; S. De Vries et al., "Local Availability of Green and Blue Space and Prevalence of Common Mental Disorders in the Netherlands," *British Journal of Psychiatry* 2 (2016): 366–72.
24. "Sierra Club Names New Environmental Justice Award after Dr. Robert Bullard," *The Planet*, August 5, 2014, https://www.sierraclub.org/planet/2014/08/sierra-club-names-new-environmental-justice-award-after-dr-robert-bullard.

Appendices

1. https://www.fda.gov/Food/IngredientsPackagingLabeling/GRAS/.
2. "Glossary of Environmental Terms," Arizona Department of Environmental Quality, https://legacy.azdeq.gov/function/help/glossary.html.

Index

Page numbers in italics indicate figures.

About the Author

Harriet A. Washington has been the Shearing Fellow at the University of Nevada's Black Mountain Institute, a research fellow in medical ethics at Harvard Medical School, a senior research scholar at the National Center for Bioethics at Tuskegee University, and a visiting scholar at DePaul University College of Law. She has also held fellowships at the Harvard T. H. Chan School of Public Health and Stanford University and is a lecturer in bioethics at Columbia University. She is the author of *Deadly Monopolies, Infectious Madness,* and *Medical Apartheid,* which won a National Book Critics Circle Award, the PEN Oakland Award, and Black Caucus of the American Library Association Nonfiction Award.